Australian   Shells

Frontispiece:
Ficus subintermedia
D'ORBIGNY, 1852
(*Photo*—Don Byrne)

Australian SHELLS

illustrating and describing 600 species of
marine gastropods found in Australian waters

B. R. WILSON PhD
Curator of Molluscs
Western Australian Museum

KEITH GILLETT FRPS, AFIAP
Fellow of the Royal Microscopical Society

CHARLES E. TUTTLE COMPANY
Rutland, Vermont & Tokyo, Japan

*This Tuttle edition is the only edition authorized for sale in
North America, South America, the Middle East, and Asia*

*Published by the Charles E. Tuttle Company, Inc.
of Rutland, Vermont and Tokyo, Japan
with editorial offices at Suido 1-chome, 2-6, Bunkyo-ku, Tokyo, Japan
by special arrangement with A. H. & A. W. Reed, Sydney, Melbourne,
Wellington, and Auckland*

© 1971 by B. R. Wilson—K. Gillett

All rights reserved

Library of Congress Catalog Card No. 70-161484

International Standard Book No. 0-8048-1015-X

*First Tuttle edition published 1972
Second printing, 1972*

PRINTED IN JAPAN

preface

Australia is unique in the world of sea shells. No country can claim to have as many different or as many beautiful kinds. From the warm tropical waters of the Great Barrier Reef to the cool, blue depths off Tasmania, a thousand forms of shells flourish along the varied shores. It is a heritage of nature at the doorstep of Australians that offers many rich rewards in collecting, observing and studying.

This beautiful book will awaken an interest in the seas of Australia. Let us hope that it will encourage Australians to protect rather than destroy their precious wildlife of the seas.

It is refreshing to have a highly trained and competent specialist, like Dr Barry Wilson, descend from his "ivory tower" to write a book for the layman. Interpreting science is a task best done by a scientist who also has the ability to write.

The beautiful color plates made from photographs by Keith Gillett should be a boon to collectors seeking to identify Australian shells and the inclusion of so many illustrations of living mollusks reflects the biological interests of both authors and makes the subject come alive on the printed page.

Marine snails, the subject of this book, are only a small part of the world of mollusks.

Novice conchologists now have a splendid book that will point the way to newer fields of interest for the legions of Australian naturalists of the future.

R. TUCKER ABBOTT
The du Pont Chair of Malacology,
Delaware Museum of Natural History

Plate 1. *Prionovolva cavanaghi* IREDALE, 1931. (*Photo*—Lin Jones).

acknowledgments:

We deeply appreciate the co-operation of the Western Australian Museum Board of Trustees and the Museum Director, Dr W.D.L. Ride, who generously allowed us to use the collections and facilities of the W. A. Museum.

To the Directors and staff of the Australian Museum, Sydney, the National Museum of Victoria, and the South Australian Museum we also extend our thanks for the use of specimens from their collections. Mrs Shirley Slack-Smith and Miss Anne Paterson of the W. A. Museum, Dr Winston Ponder of the Australian Museum and Dr Brian Smith and Mr Tom Darragh of the National Museum of Victoria, warrant special thanks for their help and advice throughout the compilation of the book. Mrs G. E. "Lally" Handley performed the onerous task of typing and retyping the manuscript and her patience and efficiency are greatly appreciated.

For loaning specimens for illustration we are extremely grateful to: Frank Abbottsmith, Mr and Mrs Wally Back, Mr and Mrs Stan Ball, Neville Coleman, Walter Deas, Aileen East, Tom Garrard, Walter Gibbins, David Greenacre, Mr and Mrs Alan Hansen, Tony Kalnins (Western Australian Shells, Perth), Lance Moore (Marine Specimens, Sydney), Mr and Mrs Mal Parker (Sea Shells Unlimited, Fremantle), Morac Seymour, Thora Whitehead and Mr and Mrs Jack Yates.

Frank Abbottsmith, Tom Garrard, Neville Coleman and Thora Whitehead also gave valuable advice concerning the families Volutidae, Terebridae, Muricidae and Conidae respectively.

For proof-reading the text we are indebted to Tom Jarrett, John Reed and John O'Reilly.

Special thanks are also due to Charles Turner for his skill and patience with the colour film processing. For assistance with the development of new electronic equipment and techniques used in photography for this book we would like to thank Ron and Terry Barlow.

Don Byrne, Neville Coleman, Walter Deas, Jack Fisher and Doug Henderson generously supplied photographs of living molluscs.

Brian Bertram is to be congratulated for his skill and perseverance in preparing the line drawings for the book.

We are indebted to our wives Greta and Ailsa for their patience and understanding during the compilation of the entire work.

Fig. 1. *Fusinus novaehollandiae* REEVE, 1848.

Plate 2. *Phalium areola* LINNAEUS, 1758. At a depth of 6 feet, Bell's Beach, Mossman, N. Qld. (*Photo*—Neville Coleman).

contents

Fig. 2. *Epitonium imperialis* SOWERBY, 1844, drawn from colour photograph of a living animal.

introduction

OF THE MAJOR GROUPS in the animal kingdom, the Mollusca is among the largest and most diverse with representatives living in the sea, in freshwater lakes and rivers, and on land. Recent estimates of the number of species of molluscs living today vary from 80,000 to over 100,000. The shores of the Australian island continent are renowned for their rich fauna of marine molluscs. No one can say exactly how many species live there but the number must be in the order of tens of thousands. This abundance is at once a blessing and a curse to the Australian shell collector. Most of us begin by collecting all kinds of shells but very soon realise that we have not the time, space, or mental energy to deal with such a vast number of specimens and species. Also, it soon becomes evident that, while it is easy to collect a large number of species at a single locality, it is quite impossible for most collectors to find names for more than a few of them.

Many groups of molluscs in Australia are very poorly known and, even for the better-known families, there is a shortage of up-to-date, illustrated, popular literature. This book is intended to foster interest in Australian shells by helping those who wish to identify specimens or to learn something of the animals which make them.

It is not possible in any single book to describe or illustrate every known species of Australian marine mollusc, nor is it possible, because of the diversity of these animals, to consider representatives of every family. In this book certain families of special interest have been selected and an attempt has been made to deal with as many as possible of the Australian marine species belonging to them. The families selected are those which include the species most prized by shell collectors or most commonly noticed by the casual wanderer on the seashore. All

of them belong to the class Gastropoda and no representatives of the other major groups of molluscs have been included.

Classification and Names of Australian Shells

Animals and plants are classified by taxonomists into groups of similar kinds, the highest-ranking group being the phylum. Molluscs belong to the phylum Mollusca. Like other phyla the Mollusca is divided into smaller and smaller groups of decreasing rank as follows: class, order, family, genus and species. There are usually subdivisions of each of these categories as well. Because of space considerations, and the fact that only one of the classes is dealt with in this book, no detailed discussion of the classification and general characters of the Mollusca is given here, but it may be helpful to list the classes.

1. CLASS MONOPLACOPHORA: simple molluscs with limpet-like shells, only two or three living species, all found in very deep water, marine.

2. CLASS AMPHINEURA: chitons or coat-of-mail shells and their relatives; several hundred species, all marine.

3. CLASS SCAPHOPODA: tusk shells; several hundred species, all marine.

4. CLASS BIVALVIA: bivalves (or pelecypods); about 15,000 species, marine and freshwater.

5. CLASS CEPHALOPODA: octopuses, squids, nautiluses and their relatives; several hundred species, all marine.

6. CLASS GASTROPODA: snails and slugs; the largest class with about 90,000 living species, marine, freshwater and terrestrial.

Plate 3. *Cymbiolacca wisemani* BRAZIER, 1870. At a depth of 8 feet, St. Crispin's Reef, off Mossman, N. Qld. (*Photo*—Neville Coleman).

9

Opinions vary on the question of how the class Gastropoda should be divided. Most authors recognize three sub-classes. The sub-class Prosobranchiata includes the majority of the marine snails and a large number of land snails. They have gills in the mantle cavity and a large proportion of them are able to seal the aperture with an operculum. Classification of the sub-class Prosobranchiata relies on details of the radula and anatomy. One useful classification recognizes three orders, the Archaeogastropoda, the Mesogastropoda, and the Neogastropoda. All the molluscs considered in this book are marine prosobranchs.

The sub-class Opisthobranchiata includes marine snails and slugs such as bubble-shells, sea-hares and nudibranchs. These forms are hermaphroditic and the shell is commonly reduced or absent. The sub-class Pulmonata includes hermaphroditic land snails and slugs which lack an operculum and gills and breathe by means of a "lung" which is actually a modified mantle cavity. Neither of these two sub-classes of the Gastropoda is dealt with here.

Once a group of animals has been subdivided into units of progressively lower rank, the units or categories are each given a name. The procedures of naming categories are called nomenclature and they are quite complex and governed by a set of rules drawn up by the International Commission on Zoological Nomenclature.

It is the convention that each kind of animal has a name comprising at least two words, the first being the generic name and the second the specific or trivial name. The generic name is always written with a capital first letter but the species name is not. Both names should be used when referring to a particular species although it is permissible to abbreviate the generic name to its first letter after the name has been written in full once.

Sometimes a subgeneric name is also used, in which case it comes after the generic name and before the species name and is enclosed in brackets. If a subspecies name is used it follows the species name. For example the technical name of the spider conch is written *Lambis chiragra*. If a subgeneric name is included it is written *Lambis (Harpago) chiragra*. The Australian subspecies is the nominate form and its full name is written *Lambis (Harpago) chiragra chiragra*, but the East African subspecies is *Lambis (Harpago) chiragra arthritica*. The abbreviation of the last name would be *L. (H.) c. arthritica*. It is not usually necessary for either the subgeneric or subspecific name to be used but sometimes it helps to do so when finer precision of meaning is required.

Another convention of taxonomy is the inclusion of the name of the person who first described the species, and the date of his publication, after the species name when it is introduced for the first time, e.g. *Lambis chiragra* LINNAEUS, 1758. This also applies for subspecies, e.g. *L. chiragra chiragra* LINNAEUS, 1758, and *L. c. arthritica* RÖDING, 1798.

Australian shell collectors often complain that scientists keep changing the names of our shells. This is true but we must keep in mind the reasons for it.

The need to change the names of animals may arise in several ways. Sometimes one species may be given two or more names by different taxonomists not familiar with each other's work. Later, when this is discovered, the name which was published first is accepted, the other is regarded as a *synonym* and is rejected. This is the "Law of priority" laid down by the International Commission.

Sometimes a new name may be introduced for a species when the same name had been used earlier for a different species of the same genus. Again the law of priority asserts that only the earlier usage must be accepted. In this case the later unacceptable usage of the name is said to be a *homonym*. The one name may be used for different species in different genera but generic names must be unique in the animal kingdom.

Name-changes of this kind are the result of inadequate study and literature research by the original authors (not always avoidable). The changes are necessary for essentially legalistic reasons and have little to do with biology. They are a nuisance to everyone. However, they are merely an unhappy aspect of a nomenclatural system designed to minimise name-changes. If the system did not exist at all, if there were no rules to determine priorities when problems arise, then the names of shells and other creatures would soon become hopelessly confused.

Other name-changes result not for legalistic reasons but because of new information, available through more advanced research techniques, showing that the old classification system must be modified. This affects generic names most seriously but it can also necessitate changes in species names because of homonymy. Revision of classification and changes in names are also made necessary by discovery of new kinds of animals. When the taxonomist attempts to fit these into the existing system he may find that this cannot be done without modification of the classification. Sometimes changes are made necessary when information from different parts of the world are brought together for the first time.

The nomenclature of the Australian Mollusca is still in a thoroughly confused state. There are not enough trained taxonomists for the mammoth task of properly classifying this great phylum, and new data and methods are continually being discovered which necessitate the revision of older works. The work of many Australian malacologists over the years has suffered from the view that the Australian marine fauna is predominantly peculiar to our shores. This view is substantially true of the marine fauna of southern Australia, but many of the species found on our northern coast have wide distributions in the Indian and Pacific Oceans. The consequence of this belief has been the introduction of a vast number of new names for Australian shells, particularly for those of our northern coasts, without proper comparison of them with known and named species from neighbouring countries. Thus much of the current taxonomic effort of Australian malacologists is devoted to the reassessment of the relationships and identity of our molluscs, and many familiar names well established in our literature are having to be discarded in favour of older names originally introduced for specimens from other places.

Although the prospect is not pleasing we must expect frequent name-changes of genera and species for many years to come. The species names used in this book are as up to date as possible but errors and oversights undoubtedly occur. Generic names have proven to be even more difficult. In some families (e.g. cowries) broad generic units have been used and the finer subdivisions commonly applied in recent years have been avoided. The original diagnoses of many of these subdivisions are so inadequate and ambiguous that final acceptance of them must await more comprehensive and definitive studies than this.

Identifying Australian Shells

Identification of specimens is one of the most interesting and important activities for the serious shell collector. It can also be most frustrating and time-consuming. There are several reasons for this. As discussed in the previous section, there are many taxonomic problems concerning molluscs which have not yet been resolved because of lack of data or lack of study. In such instances, authors of identification handbooks such as this can only make calculated guesses, and so opinions may vary from one book to another. Few books are able to give adequate descriptions and illustrations of all the species to be found in the regions to which they apply. Some species of shells are so variable in shape, colour or sculpture that it is not possible to describe all the variations. Such variations may be the result of environmental influences or genetic variation within the species, or variation between juvenile and adult shells of the same species.

There are some instances in which accurate identification cannot be made on the basis of shell characters alone, and it becomes necessary to look at the anatomy of the animals. However, many collectors mistakenly believe that examination of the radula by a specialist will solve all difficult problems concerning the identification of gastropods. Unfortunately this is not true, for although the characteristics of the radula may provide helpful or even positive answers in some species, it can be just as variable as any other structure.

For reasons such as these, shell collectors should not expect to be able to identify all the shells they find. Through perseverance and experience the beginner will learn to use the books and articles available for his area and be able to find names for most of his specimens, but he will soon reach a stage when his problems are beyond the scope of available books.

He will then find that membership of a shell club will be of great help through discussion of his problems with other collectors. Most of Australia's State museums now employ professional staff specialising in molluscs, and are able to help collectors with identification problems. However, collectors should keep in mind the fact that the museums' shell departments are still in a developmental stage and that their staff have to compromise between time spent in answering public enquiries and time spent on the museum reference collections and doing the basic research which, in the long run, will make an adequate identification service possible.

A list of books, checklists and scientific journals relevant to Australian shells is given at the end of this section. Some of these publications deal specifically with Australian shells, others deal with shells of other parts of the Indo-West Pacific region but contain information about Australian species.

The Gastropod Animal

Almost everyone will have seen a soft-bodied garden snail, foot extended and eye stalks erect, crawling on the plants and leaving behind trails of shiny mucus. The bodies of most marine gastropods are basically similar to those of garden snails although there are important internal differences.

The two major regions of the gastropod body are the muscular head-foot complex and the non-muscular visceral mass. The foot of gastropods is a mucous-secreting muscular thickening on the under-surface of the body, responsible for the gliding or creeping form of locomotion typical of these creatures. At its front end is the head where the mouth (often situated in the end of a long extensible proboscis), the eye tentacles, and the simple brain are located. Because the head and foot form a more or less continuous muscular structure, they are referred to as the "head-foot complex". On top of the foot and behind the head is the visceral mass which contains such vital organs as the stomach, intestines, liver, kidney, heart and reproductive organs.

In the majority of gastropods the visceral mass is coiled into spiral whorls and is protected by a similarly coiled shell secreted by the mantle. Usually, the foot and head can also be drawn within the protection of the last or body whorl of the shell, and often the

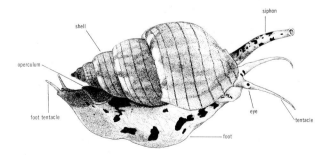

Fig. 3. *Nassarius particeps* HEDLEY, 1915, showing the major external features of the body. Note especially the position of the operculum in the crawling snail.

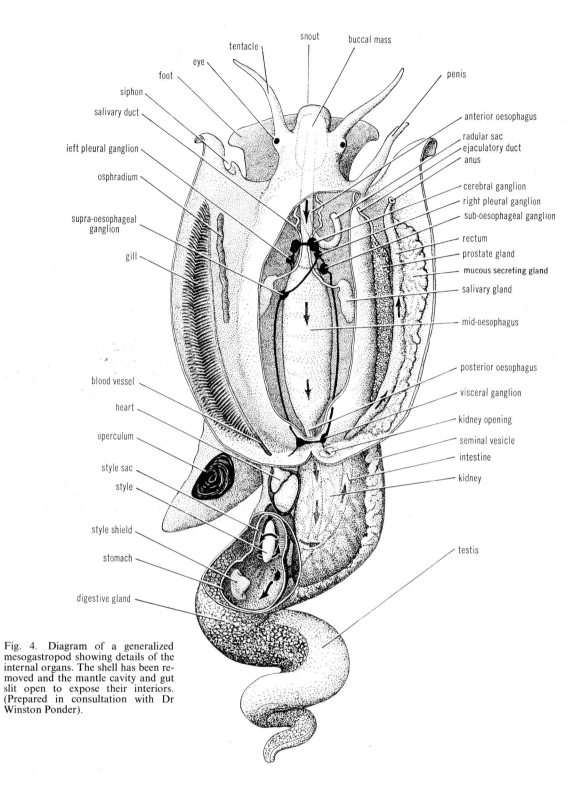

tentacle

snout

buccal mass

eye

penis

foot

siphon

salivary duct

anterior oesophagus

left pleural ganglion

radular sac

ejaculatory duct

anus

osphradium

cerebral ganglion

right pleural ganglion

sub-oesophageal ganglion

supra-oesophageal
ganglion

rectum

prostate gland

mucous secreting gland

gill

salivary gland

mid-oesophagus

posterior oesophagus

blood vessel

visceral ganglion

heart

kidney opening

operculum

seminal vesicle

style sac

intestine

style

kidney

style shield

stomach

testis

digestive gland

Fig. 4. Diagram of a generalized mesogastropod showing details of the internal organs. The shell has been removed and the mantle cavity and gut slit open to expose their interiors. (Prepared in consultation with Dr Winston Ponder).

aperture is sealed by an operculum or "door" which is attached to the back of the foot.

Draped over the visceral mass is a very important organ called the mantle. It is a thin "skin-like" membrane which adheres to the sides and top of the visceral mass but hangs free like a skirt around the lower edges. At the front, just behind the head in prosobranchs and pulmonates, there is a deep cavity between the mantle and the visceral mass. This is the

mantle cavity which contains the gills (ctenidia), large mucous-secreting glands and sensory organs called the osphradia sensitive to the chemical composition of the water. It also contains the external openings for the alimentary, excretory and reproductive organs.

Water is pumped through the mantle cavity by means of the beating action of countless tiny hair-like structures called cilia and by muscular contractions.

⇧

Plate 4. *Conus textile* Linnaeus, 1758, with egg capsules at a depth of 6 feet, Langford Island, Qld.
(*Photo*—Neville Coleman).

Plate 5. *Cypraea (Zoila) friendii* Gray, 1831, in a typical situation on the side of a sponge at a depth of 60 feet, off ⇨ Dunsborough, Geographe Bay, W. A. (*Photo*—Barry Wilson).

The water brings oxygen to the gills and carries away respiratory, digestive and excretory wastes, and also eggs and sperm in those species which reproduce by external fertilization. In more advanced gastropods the water is drawn in through a tubular fold of the mantle edge called the anterior or incurrent siphon. The outgoing water may be channelled through another but usually shorter mantle fold called the posterior or excurrent siphon. Often the siphons are supported by projecting folds of the shell outer lip which are also referred to as the anterior and posterior siphons or siphon canals.

In all classes except the Gastropoda, the adult mantle cavity is at the rear or sides of the body. A characteristic of gastropods is that, during early development, the visceral mass twists around in relation to the head-foot, so that the mantle cavity is brought into an anterior position. This process is called torsion. The effects of torsion have had far-reaching results on the evolution and diversification of the gastropods. In the prosobranchs and pulmonates the mantle cavity remains in its anterior position, but in the opisthobranchs the reduction or loss of the shell is accompanied by a tendency toward secondary de-torsion so that the mantle cavity lies on the right-hand side or is lost altogether in the case of shell-less nudibranchs.

Feeding

The earliest molluscs are believed to have fed upon fine organic particles gathered from the sea floor. This was aided by a rasp-like structure called the radula situated in the floor of the mouth. The radula is still an important feeding organ in most modern gastropods but it has been modified in a variety of ways for obtaining food of many different kinds.

Archaeogastropods such as abalones, limpets, trochids and turbans use their radulae to rasp particles of algal material from rocks or seaweeds. In these animals the radula is simple in structure, consisting of a flexible chitinous strip bearing transverse rows of pointed cusps or "teeth". It is rubbed over the food by the action of the muscles which hold it in position in the mouth.

In the mesogastropods the number of cusps per row is much reduced and the cusps are modified to perform special functions such as biting, or drilling holes in the shells of the prey. Mesogastropods feed on a variety of substances ranging from algae and the polyps of coelenterates or other sedentary animals, to more active prey such as worms, other molluscs or even spiny echinoderms. Some mesogastropods are parasites, especially on echinoderm hosts. A few filter plankton or detrital material from the water.

Neogastropods, such as the murex shells, volutes, cones and augers are exclusively predators. They are active snails and prey on other invertebrates or, in the case of some cones, on fish. The radular teeth in these forms are reduced to three or one per row and they are highly specialised for killing the prey.

The feeding habits of many molluscs are still unknown and the amateur shell collector can contribute much useful information on this subject from field and aquarium observations.

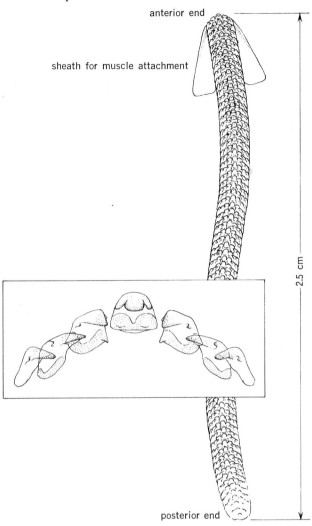

Fig. 5. Diagram of a gastropod radula, *Cypraea (Zoila) friendii* GRAY, 1831, a typical taenioglossate radula with 7 teeth per row. Teeth at the anterior end are worn off during feeding. New teeth are formed at the posterior end.
Insert—Enlargement of a single row of teeth.

Reproduction and Development

Molluscs reproduce in a great variety of ways. Many marine species have simple reproductive systems consisting only of a gonad where the eggs or sperms are produced and a duct from there to the exterior. At spawning time the eggs or sperm are simply ejected into the water through the duct and fertilization is said to be "external". In these species vast numbers of small eggs are produced by each female, the adults tend to be gregarious, and the embryos develop as they drift and swim in the water. They pass through two larval stages, the trochophore and the veliger. Veligers have a swimming organ known as the velum, which is lost in later life. They also have a tiny shell which, in the case of gastropods, may be visible still at the tip of the spire of adult shells and is known as the protoconch. These free-swimming or pelagic larvae feed on phytoplankton. When larval development is completed the veligers sink to the bottom and settle on a suitable substrate. Often they tend to favour places where there are adults of their species so that the suitability of the substrate is guaranteed. After settlement, changes take place in feeding, physiology and behaviour, and the animals take on the adult way of life. This kind of reproduction and development occurs in abalones, limpets and some other archaeogastropods.

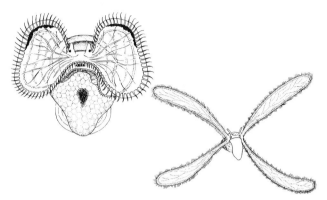

Fig. 6.   Free-swimming veliger larvae of two European gastropods, showing in each case a large velum.
Left: *Littorina littorea* (x30).
Right: *Simnia patula* (x13).

Most gastropods have more complicated reproduction systems with accessory organs, glands and ducts. The male and female animals copulate and sperm are deposited within the reproductive tract of the female. Fertilization is said to be "internal". After fertilization the female lays her eggs in protective capsules made of glandular secretions and early development of the embryos takes place within these protective walls. The mother also includes supplies of yolk within the eggs to nourish the developing embryos. Females of molluscs which reproduce in this way lay relatively few, large eggs. Sometimes there are several hundred eggs in each capsule and all of these become viable larvae. In other species only one or two of the eggs develop and the others, the "nurse eggs", are used as food. In most mesogastropod families the larvae hatch from their capsules in the free-swimming veliger stage. However some mesogastropods and most of the neogastropods have direct development, i.e. the entire development takes place within the capsules and the young hatch as tiny crawling snails (see Fig. 33, page 148).

The variations on these basic reproductive themes are countless, even within the Gastropoda. Some species are hermaphroditic, some change sex once or several times during their life, some brood their young in brood pouches, others lay their eggs in the bodies of other animals. Wherever possible a brief summary of reproduction and development is given in the introductory sections for each family in this book.

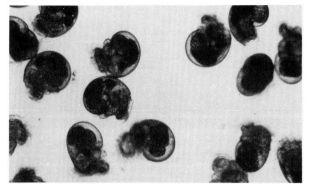

Fig. 7.   Photomicrograph (x40) of veliger stage embryos of *Dicathais orbita* GMELIN, 1791, as they hatch from their egg capsule. (*Photo*—Keith Gillett).

The Gastropod Shell

Composition

The shell of a mollusc is composed of crystals of calcium carbonate (lime) deposited in a matrix of an organic proteinaceous substance called conchiolin. There is usually an outer covering, the periostracum, composed of another organic protein material. The tissue responsible for shell formation is the mantle. Calcium carbonate is deposited in any of three crystalline forms, calcite, aragonite, or vaterite. The structural arrangement of the crystals varies. There is usually an outer, prismatic layer in which the crystals are arranged as vertical columns, and an inner layer, sometimes nacreous, formed of thin, laminated sheets of crystals. Calcium is taken into the body either in the food or directly by absorption from the environment. Within the body fluids it combines with carbonate, but formation of the calcium carbonate crystals takes place in a thin layer of fluid lying between the outer surface of the mantle and the inner surface of the shell.

Shape, Sculpture and Ornament

The shell of shell-bearing gastropods usually takes the form of a tapering tube, closed at its narrow end and open at the other, but the variations on this basic structure are many. In most species the tube is coiled into a spiral. The shells of adult limpets are not coiled but have a simple conical shape. There is no shell at all in many of the marine and terrestrial slugs. Special terms are used to describe the shapes of gastropod shells and the most commonly used ones are illustrated in Fig. 8. The parts of the shell also have names, and the more important of these are shown in Fig. 9.

Often the shape and sculpture of a particular shell may be explained in functional terms if sufficient is known about the life of the animal which made it. For instance, the long anterior siphon canal of many gastropods provides support and protection for a soft extensible fold of the mantle which controls the flow of water to the gills. The longitudinal varices or ridges of such shells as the cymatiums and murexes strengthen the shell and are formed as thickenings during periods when there is no growth. Sometimes spines are formed from the varices which protect or camouflage the animal. The long aperture of cowry shells is related to the way water flows through the gill chamber. Some gastropods which prey on bivalves have specially shaped or thickened outer lips or outer lip spines for use in prising apart the valves of the prey. However, in many cases the function of ornament or fine sculpture may be not at all obvious and we have to fall back on the assumption that it is genetically involved with other biologically important features of the animal.

Fig. 8.   Some terms referring to shell shape.

Colour and Patterns

Colour patterns of shells are produced by deposition of pigments during formation of the outer shell layer by the edge of the mantle. Spots, lines, circles, bands and other marks are formed by movement of the pigment-secreting cells in the mantle edge as the

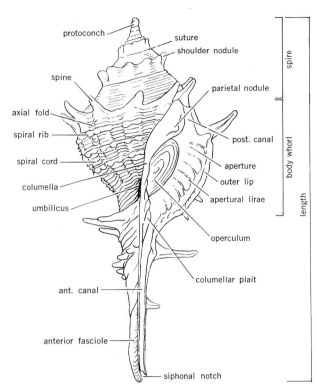

Fig. 9.   Diagram of a hypothetical gastropod shell showing many of the structures used in descriptions.

shell formation process goes on. Each species has its own specific colour pattern, or range of patterns, and these are often used in the description and identification of species. However we know very little about the function of the intricate patterns and beautiful colours which make many shells so attractive to our eye. It is possible that in some cases they may help the animals recognise each other, but this cannot explain the development of intricate pigmentation of shells such as the cones, which have a thick periostracum rendering the pattern on the surface of the shell more-or-less invisible. It has been suggested that the pigments may be unwanted products of the metabolism of the mollusc's body and that their deposition in the shell layers is simply a convenient way of disposing of them. This is one of the many unresolved problems concerning the natural history of molluscs.

Growth and age

As the animal's body grows so the shell size must increase to accommodate it. The volume of the shell is increased by addition of new material to the edge of the outer lip through deposition of calcium carbonate from the mantle edge. At the same time the shell is also thickened and strengthened by additions to its inner surface by secretions from the outer surface of the mantle. Each phase of deposition of shell material on the lip edge is marked by a growth line. In some species growth proceeds in stops and starts and the periods when it stops are marked by thickened varices or prominent "growth rings". The geometric precision of angle, curvature and dimension during growth of the shell is truly an incredible phenomenon, for the final result is a shape unique, or almost so, for each species.

Initially growth may be very fast, but the rate declines as time goes on. Many molluscs have no genetically determined maximum age or size but continue growing at a very slow rate until they die of disease or misadventure. Others, like cowries and strombs, stop growing when they reach sexual maturity. The lip is then curled inward (in cowries) or outward (in strombs) and the shell then grows thicker but not longer or more voluminous.

Maximum age of molluscs is very difficult to determine. Most species live for a few years but some live for only one year and others may live for at least fifty years. The only reliable way of telling the age of molluscs is to keep individuals under observation. It is sometimes possible to make reasonable estimates by counting growth rings on the shell, although, before this can be done, it is necessary to show what period of time each growth ring represents. This can be determined only by intensive biological study of the species concerned.

Geographic Distribution and affinities of Australian Shells

Naturalists who have had the good fortune to compare shells from the southern (temperate) and northern (tropical) coasts of Australia will observe that there are very few species common to both areas. In fact these two areas are generally regarded as two distinct marine faunal regions. They overlap on the western and eastern sides of the continent. These regions and other matters concerning the distribution and relationships (i.e. the zoogeography) of Australian molluscs are discussed below.

The Northern Australian Region

The majority of shallow-water marine animals, including the molluscs, found along the north coast of Australia also occur in the seas around the islands of the Indo-Malaysian Archipelago and the south-east Asian mainland, while many are widely distributed throughout the tropical Indian and Western Pacific Oceans. So the northern coast of Australia is generally regarded as part of the great Indo-West Pacific faunal region. Nevertheless, there are a number of species endemic to northern Australia. There are also some differences between the faunas of northern Western Australia, and the east coast of Queensland. Differences have also been observed between the faunas of the offshore Great Barrier Reef and the mainland shores of Queensland.

Some examples of molluscs endemic to northern Western Australia are *Cypraea decipiens*, *Amoria praetexta* and *Conus victoriae*. Examples of endemic Queensland species are *Cypraea xanthodon*, *Cymbiolacca pulchra* and *Conus sculleti*. There are a few cases of closely related species pairs, apparently separated relatively recently on either side of the Torres Strait, e.g. *Volutoconus grossi* of Queensland and *V. hargreavesi* of northern Western Australia.

There are also many examples of subspecies pairs on either side of the north coast, e.g. *Cypraea subviridis subviridis* in the east and *C. subviridis dorsalis* in the west.

The Southern Australian Region

The temperate marine fauna of southern Australia is unique. In some measure it seems to have been derived from the fauna which once lived in the central Indo-West Pacific Region for it has many generic and subgeneric similarities with the fauna living in that region today, and little in common with the faunas of South Africa and South America. But the origins of some of the more ancient elements in the marine fauna of southern Australia remain a mystery to us.

There are many endemic generic and subgeneric groups in the Southern Australian Region, especially off the south-eastern corner of the continent (e.g. *Notovoluta, Notocypraea, Austroharpa*). Fossil evidence indicates that many of these are relics of groups which have inhabited this temperate region since the early Tertiary (i.e. as much as 50 million years ago). Some of them have no known close fossil or living relatives, but others are related to fossil forms found in the early Tertiary rocks of Asia and southern Europe. Thus, it seems that a distinctive molluscan fauna has been present in southern Australia for a very long time. However, there is evidence that in later Tertiary times there were invasions of species from the tropical Indo-West Pacific Region. These invaders have left modern descendants which have close relationships with living species in the north (e.g. *Oliva australis, Melo miltonis*).

The Eastern and Western Overlap Zones

On the mid-west and mid-east coasts of Australia there are wide overlap or transition zones between the northern (tropical) and southern (temperate) faunas. In Western Australia some northern species (e.g. *Cypraea caputserpentis*) extend as far south as Cape Leeuwin and some southern species (e.g. *Melaraphe unifasciata*) extend as far north as North-West Cape. Near Fremantle the fauna of marine molluscs is a mixture of northern and southern species although the southern species predominate. A similar situation prevails on the east coast where the overlap zone lies in an area approximated by the northern and southern state boundaries of New South Wales.

Both overlap zones possess a number of endemic species. For example, in Western Australia *Cypraea rosselli*, *Aulicina nivosa* and *Conus nodulosus* are found only in the mid-west overlap zone; in the eastern Australian overlap zone *Lyreneta laseroni*, *Cymbiolena magnifica* and *Conus papilliferus* are endemic species.

Marine Faunal Provinces in Australia

Not deterred by the problems associated with classification and nomenclature of marine molluscs and other creatures, many Australian taxonomists

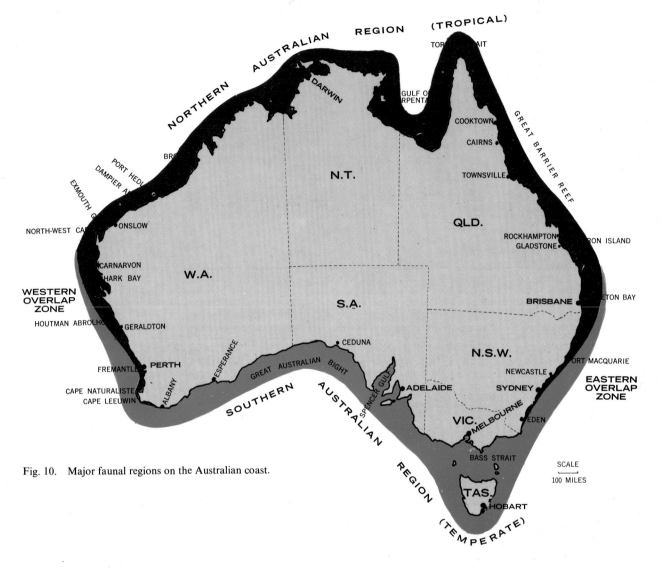

Fig. 10.   Major faunal regions on the Australian coast.

have attempted to classify our coastline into numerous faunal provinces. Concepts of the meaning and significance of faunal provinces vary, but usually the term is taken to mean an area or province containing a fauna which includes a high proportion of endemic species, i.e. species not found elsewhere. Such divisions are usually based on lists of species of each province, but our knowledge of the Australian marine fauna is, as yet, very imprecise. Consequently, views on the number and limits of faunal provinces vary, often depending on the group of animals in which the taxonomist specialises. Because readers may come across such terms as "Dampierian species", "Flindersian species" or "Peronian species" in other books or scientific papers on Australian molluscs, the faunal provinces commonly adopted by Australian malacologists are listed here. Readers should be warned, however, not to place too much reliance on these very imprecise terms. They can be misleading for they place too much emphasis on the differences between the faunas of the provinces and tend to obscure the affinities they have with each other and, in the case of the northern provinces, with the faunas of other parts of the world.

1. **Northern Australian Region**
    (i) Dampierian Province—northern Western Australia and the Northern Territory.
    (ii) Solanderian Province—the offshore islands and reefs of the Great Barrier Reef.
    (iii) Banksian Province—the mainland shores of Queensland. (Some biogeographers include the Northern Territory).

2. **Southern Australian Region**
    (i) Flindersian Province—southern Western Australia, South Australia and western Victoria.
    (ii) Maugean Province—Tasmania and southern and eastern Victoria.

3. **Eastern Overlap Zone**
    (i) Peronian Province—New South Wales and extending slightly into eastern Victoria and southern Queensland.

4. **Western Overlap Zone**
    (i) No province named—extends from North-West Cape to Cape Leeuwin, Western Australia.

Plate 6. *Calpurnus verrucosus* LINNAEUS, 1758, with eggs, in its natural habitat on the alcyonarian, *Lobophytum pauciflorum* EHRENBERG, 1833, at a depth of 6 feet, Langford Island, Qld.   (*Photo*—Neville Coleman).

Relationship of the
Australian and New Zealand Faunas

For many millions of years the continental shelf areas surrounding the islands of New Zealand have been separated from those of Australia by more than 900 miles of ocean reaching depths of more than 2,000 fathoms. As might be expected then, the marine faunas of these two countries are distinct. Yet there are a number of molluscs common to both Australia and New Zealand, e.g. *Phalium labiatum, Tonna tetracotula, Niotha pyrrhus, Agnewia tritoniformis, Nerita atramentosa.* Most of these species have long-lived pelagic larval stages capable of being carried great distances by ocean currents and it is believed that they were introduced to New Zealand as immigrants from Australia transported by eastward flowing currents. A few species may have been transported as adults attached to floating logs. Introduction of New Zealand molluscs to Australia seems to have been less frequent, presumably because of the direction of the ocean currents.

The fact that, in these particular immigrant species, the daughter populations in New Zealand have not diverged from the ancestral Australian stock, suggests either that the immigration was a relatively recent event, or that their larvae continue to flow into New Zealand waters fairly regularly. However, there are some other New Zealand populations, believed to be derived from Australian ancestors, which have diverged to the extent that they are now recognised as distinct species or subspecies (e.g. *Charonia capax* and *Dicathais scalaris* of New Zealand, derived from *C. rubicunda* and *D. orbita* of Australia). There is also fossil evidence that immigration of Australian molluscs into New Zealand has been going on intermittently for several million years.

Conservation of Australian Shells

Throughout the world mounting concern is being expressed by scientists and laymen about the destruction and pollution of the earth's environments. Most of the obvious damage, and so most of the protests, concern terrestrial environments but there is good reason to fear for the wellbeing of the oceans too. Marine bays and estuaries all over the world are being used as "rubbish bins" for man's wastes of all kinds. Some are also being seriously affected by silt, insecticides and fertilisers carried down by rivers from developing agricultural areas. Some of this pollution may be unavoidable but much could be controlled if we cared enough about it. The result today in parts of the coasts of Europe and the United States is fouled estuaries and bays which have become biological deserts. They no longer serve as nursery grounds for commercial fisheries, nor are they satisfactory recreation places for man's enjoyment.

One result of our increasing affluence and leisure has been the recent extraordinary increase in the popularity of shell collecting as a hobby. Shell collectors are by nature the kind of people who enjoy the freshness and peacefulness of the seashore and delight in observing the multitude of diverse creatures which dwell there. Yet we also have within us still a little of the primitive hunter and more than our share of human acquisitiveness. These traits have put us in a glass house from which we should hesitate to throw stones at industrialists and agriculturalists about pollution and conservation. The actions of many shell collectors are far from beyond reproach with regard to this matter.

Irresponsible collecting of all kinds of marine animals, especially shells, is causing tremendous damage to the animal communities of Australian beaches and reefs. Bags full of indiscriminately collected shells are taken home, the best specimens are picked out for the collection, for exchange or for sale, and the rest are buried in the garden. Reefs are virtually turned upside down by shell hunters and the eggs, the food and the shelter of the reef animals are destroyed. Even now there are few places reasonably accessible to Australian cities and towns where the marine naturalist, teacher or student can observe communities of seashore animals in an unaltered state. In parts of California U.S.A. the situation is so bad that teachers at some seaboard schools and universities have to import specimens from other areas or take their students to places hundreds of miles away before they are able to look at the living animals instead of only seeing pictures of them in books.

Another aspect of this problem is that selective "fishing" of certain key species in an animal community may interfere with the integration of the whole community (the ecosystem) with drastic and unforeseen results. For example, excessive "fishing" of the giant trumpet shell *Charonia tritonis* by shell collectors and professional shell dealers has recently been proposed as a possible cause of the disastrous plague of the Crown of Thorns starfish *(Acanthaster planci)* in Queensland and other parts of the Indian and Pacific Oceans.

Over the last few years the Crown of Thorns, which was once uncommon, has increased its numbers to plague proportions at many localities. It feeds on coral by sucking out the living tissue and leaves behind the dead coral skeleton which soon erodes and crumbles away. Plagues of this starfish have moved through vast areas of coral reef and left them as algal-covered rubble which will take many years to regenerate.

A great deal of money has been spent on the study of this problem in Australia and in U.S. Pacific Territories. The cause of the plague is still not understood, but it is known that the trumpet shell is one of the few animals which eats the Crown of Thorns starfish. Field experiments on the appetite of the trumpet shell have shown that this predator may play a large part in keeping the starfish numbers down to a normal level by feeding on the young starfish before they reach maturity.

Those who object to the triton theory as an explanation of the Crown of Thorns plague point out that *Charonia tritonis* has never been a common shell on the Great Barrier Reef and that it is unfair to lay the blame at the door of shell collectors when few collectors have ever been lucky enough to find a specimen. It may yet turn out that falling numbers of some other predator not yet identified may be the cause of the problem. Whatever the final answer will be there are sufficient grounds to advocate that *Charonia tritonis* should be given full protection (as has been done in Queensland) as an ally in helping protect our coral reef, even though over-fishing of this species may not be the basic cause of the problem.

The problem of conservation in the sea is not quite the same as it is on land. Populations of most marine animals are extremely resilient and recover quickly after disaster. Also, the majority of seashore animals live in deeper water beyond the reach of most collectors, as well as in the shallows. Thus, if a reef or sand cay is "fished out" it will probably be repopulated from deeper water if left alone for a few years (provided it is free from pollution and other changes). Exceptions to this are the species which are able to live only in the intertidal zone or in estuaries or sheltered bays. But in general the problem is not that marine animals may become extinct: a more pressing aspect is that if all accessible shores are reduced to barren biological deserts, and kept that way by continued collecting, we shall all be denied the pleasures of observing seashore animals and communities in a natural state. Furthermore, opportunities for scientific research and student training will be seriously affected.

Responsible collectors should observe the following rules:

    (i) Do not collect more specimens than you need. Select one or two good-quality specimens and leave the rest to breed. If you need extra specimens for exchange use restraint and don't take so many that the majority sit in a box at home for years waiting until you hear of someone who might want one of them.

    (ii) Do not interfere with females associated with eggs.

    (iii) When you overturn stones to look for shells be careful to replace them as you found them.

There are other ways too in which shell collectors can contribute to the cause of conservation of marine life. It is important that individuals make themselves well informed on the subject by reading and discussion. Shell clubs can encourage this by organising talks and study groups. By joining local and national societies dedicated to the cause of conservation, individuals and clubs can add their weight to pressures being applied on governments and authorities concerning better anti-pollution and conservation measures. For example, in recent years there has been increasing interest in this country in the creation of marine national parks and fauna reserves where people are allowed to look at but not remove any specimens. There are many legal and administrative

Plate 7. The underside of stones and coral boulders like this are often covered with a thick growth of marine organisms such as sponges, sea-squirts, bryozoans and alcyonarians, which provide food and shelter for molluscs and other animals. This growth is killed when the stones or corals are left overturned. Location, Heron Island (Capricorn Group).
(*Photo*—Keith Gillett).

problems involved, but worthwhile results could be achieved if the need is forcefully demonstrated to our politicians.

Every naturalist-shell collector with more than a few years' experience can tell of major changes in the marine fauna of his area through pollution, over-collecting, or other human activities. The problem is serious and urgent. All shell collectors should consider their own collecting activities carefully and take an active interest in matters concerning the conservation of living molluscs and other marine creatures. Drawers full of beautifully cleaned and labelled shells will be a poor excuse when, in future years, our children want to know why they can't see things like that living on the seashore.

Names, Dates, Distribution, Abbreviations, Illustrations

Certain space-saving procedures and some departures from customary procedures have been adopted in this book and it seems appropriate that an explanation of them should be given here.

1. Scientific names versus "common" names. In natural history books it is common practice for common or vernacular names to be used for each kind of animal. There are clear advantages in doing this with such animals as birds or mammals but the value of this practice with molluscs is debatable. In any case, we have not been able to follow this custom because there are no common names for the majority of Australian species of mollusc. We see little point in inventing perhaps several hundred new common names for the species illustrated in this book. However, we have used common names for the families. Most young children and adults interested in shells simply because they are beautiful, will be satisfied to know that a specimen is an olive, or a cowry, or a cone. Serious collectors wanting to know precise names of species for the purpose of correspondence or exchange will need to learn the scientific ones, for they are the names used in nearly all shell books, catalogues and shell club magazines.

2. Authors' names and publication dates. We have already noted the convention of writing the name of the person who described a species, as well as the date of publication, after the name of the species when it is used in a book or article for the first time. It is also conventional that both the author's name and the date of publication should be enclosed within brackets if the species is being placed in a genus different from that used by the original author. We have not followed this convention; there seems little advantage in doing so as so few of our Australian molluscs are still placed in the original genera.

3. Species distribution. Readers may notice that the distribution limits given for many species seem to be more precise for Western Australia than they are for the eastern States. The eastern coast is divided by State boundaries which can be used as reference points but the western third of the continent has no such political divisions. In the west such terms as southern, central or northern Western Australia are not adequate and it is necessary to refer to towns or major bays, capes or islands.

The distribution data used in this book are drawn mainly from museum collections. It is realised that many species may be found beyond the range indicated for them here. It is hoped that collectors who can correct us on these matters will not hesitate to do so by writing to us or sending specimens to one of the State museums.

In order to avoid confusion about our distribution statements we have quoted them in a clockwise direction around Australia. Thus, "Fremantle, W.A. to northern N.S.W." means that the species ranges across the north of Australia, while "northern N.S.W. to Fremantle, W.A." means that the species ranges around the southern coast. When a species lives beyond the shores of Australia as, for example, do many of our northern shells, the overseas distribution is given first and is followed by the Australian range, e.g. "Indo-West Pacific; Pt. Cloates, W.A. to southern Qld."

4. Abbreviations. Some readers of this book may not be familiar with the abbreviations for the States commonly used in this country. For example, "S.A." could mean South Africa, South America or South Australia. For this reason we list the following abbreviations:

W.A.   Western Australia
S.A.   South Australia
Qld.   Queensland
N.S.W. New South Wales
Vic.   Victoria
Tas.   Tasmania
N.T.   Northern Territory.

5. The specimens illustrated. Gathering together the specimens illustrated in this book has not been an easy task. A large proportion of them has come from the collections of the Western Australian Museum, but many have been loaned by other museums and by private collectors. Of course it is highly desirable that a book on Australian shells should illustrate Australian specimens but we have not always succeeded in obtaining good-quality specimens from Australian localities. In these cases we have substituted non-Australian specimens rather than leave the species out. We regret this shortcoming and warn the reader against using these illustrations for any scientific purpose other than identification of specimens.

6. Which way up? A curious fact about shells and shell books is that specimens are nearly always illustrated upside down, i.e. contrary to the convention usually adopted when illustrating animals of having the anterior end at the top. For some reason most people find that shells "look right" in the upside down position. We have also illustrated our specimens in this position. Although this does seem to be more pleasing aesthetically it may cause some confusion when readers try to relate the text descriptions to the pictures. To begin with, terms like "above" and "below" are difficult to interpret. In most cases we have avoided such terms and have used instead the terms anterior or posterior which of course relate to the animal's anterior and posterior ends and not to the top or bottom of the illustration. Except in a few problematical cases, the spire of a gastropod shell is at the posterior end and the anterior siphon canal, naturally enough, is at the anterior end.

Then there is the question of the left and right sides. It is another convention that these terms relate to the animal's body when it is viewed from the dorsal side with the anterior end at the top. Thus, when an upside down shell is illustrated from the ventral aspect (apertural view) the left hand side is still on the left hand side of the picture, but when the illustration is from the dorsal aspect the left hand side of the animal is on the right hand side of the picture and vice versa.

We hope that readers will not be distracted by these complications. Perhaps the matter can be solved by using the terms "port" and "starboard", the solution that seamen have adopted to avoid confusion aboard their boats?

Reference Books for Australian Marine Shells:

ABBOTT, R. T. (1962)
Sea Shells of the World. A Golden Nature Guide.
Golden Press.
A good general guide to shell groups, but only a few Australian shells are illustrated.

ALLAN, J. (1959)
Australian Shells (revised edition).
Georgian House, Melbourne.
A standard comprehensive book.

ALLAN, J. (1956)
Cowry Shells of World Seas.
Georgian House, Melbourne.

BENNETT, I. (1967)
The Fringe of the Sea.
Rigby, Adelaide.
Very interesting reference book for animals and plants found on the seashores.

CERNOHORSKY, W. O. (1967)
Marine Shells of the Pacific.
Pacific Publications, Sydney.
A good reference book for the more popular shell groups of the Pacific (Cowries, Cones, etc.), some of which are found in northern Australia.

COTTON, B. C. and GODFREY, F. K. (1940)
The Molluscs of South Australia, Part II, Scaphopoda, Cephalopoda, Aplacophora and Crepipoda.
Government Printer, Adelaide.

COTTON, B. C. (1959)
South Australian Mollusca—Archaeogastropoda.
Government Printer, Adelaide.
A comprehensive account of the Southern Australian gastropods of the more primitive groups.

COTTON, B. C. (1961)
The Australian Mollusca—Pelecypoda.
Hawes, Government Printer, Adelaide.
A comprehensive account of Southern Australian bivalves.

COTTON, B. C. (1963)
South Australian Shells.
The South Australian Museum.
A useful small booklet containing illustrations and brief descriptions of a few South Australian species.

DAKIN, W., BENNETT, I. and POPE, E. (1960)
Australian Seashores (rev. ed.).
Angus & Robertson, Sydney and London.
Helps in the identification of common seashore animals. A good guide to seashore ecology.

GILLETT, K. (1968)
The Australian Great Barrier Reef in Colour.
A. H. & A. W. Reed, Sydney.
A useful book on tropical fauna including colour photographs of shells.

GILLETT, K. and McNEILL, F. (1962)
The Great Barrier Reef and Adjacent Islands (rev. ed.).
Coral Press, Sydney.
General book on fauna, both terrestrial and marine, of tropical islands.

GILLETT, K. and YALDWYN, J. C. (1969)
Australian Seashores in Colour.
A. H. & A. W. Reed, Sydney.
An interesting book on temperate and tropical fauna found on the seashores including colour photographs of shells.

HABE, T. (1964)
Shells of the Western Pacific in Color. Vol. II.
Hoikusha Publishing Co. Osaka.
Useful for identification of tropical Australian species of gastropods, bivalves and other molluscs.

HODGKIN, E. P. et al. (1966)
The Shelled Gastropoda of South Western Australia.
WA. Naturalists' Club Handbook No. 9.
For the identification of most common gastropods as far north as Shark Bay; well illustrated.

KIRA, T. (1962)
Shells of the Western Pacific in Color. Vol. I.
Hoikusha Publishing Co., Osaka.
Deals with those groups not covered by Habe (1964).

MACPHERSON, J. H. and GABRIEL, C. J. (1962)
Marine Molluscs of Victoria.
(Vic. National Museum Handbook No. 2).
Melbourne University Press.
Valuable for identification of all groups of molluscs from the southern half of Australia.

McMICHAEL, D. (1960)
Shells of the Australian Sea Shore.
Jacaranda Press, Brisbane.
Covers common northern and southern species, but mostly those from the eastern coast of Australia.

MARSH, T. and RIPPINGALE, O. H. (1964)
Cone Shells of the World.
Jacaranda Press, Brisbane.

MAY, W. L. (1923)
An Illustrated Index of Tasmanian Shells.
Government Printer, Hobart.

MAY, W. L. (1958)
An Illustrated Index of Tasmanian Shells (revised edition by J. H. Macpherson).
Government Printer, Hobart.

MELVIN, A. G. (1966)
Sea Shells of the World.
Paul Flesch, Melbourne and Sydney.
Many photographs of shells, with values and descriptions.

RIPPINGALE, O. H. and McMICHAEL, D. F. (1961)
Queensland and Great Barrier Reef Shells.
Jacaranda Press, Brisbane.
Illustrated with colour drawings.

TINKER, S. W. (1958)
Pacific Sea Shells (rev. ed.).
Charles Tuttle, Rutland Vermont and Tokyo.
Shells of Hawaii and the South Pacific Ocean, some ranging to northern Australia.

WAGNER, R. J. L. and ABBOTT, R. T.(1967) eds.
Van Nostrand's Standard Catalog of Shells. Second edition.
Van Nostrand Co., Princeton, Toronto, London and Melbourne.
Useful as annotated list of the species in many families and includes values.

Scientific Journals Referring to Australian Shells

Indo-Pacific Mollusca.
Edited by R. Tucker Abbott.
Previously published by the Department of Mollusks, Academy of Natural Sciences of Philadelphia. Now published by the Department of Mollusks, Delaware Museum of Natural History, Greenville, Delaware, 19807, U.S.A.
Contains world-wide revisions of families or genera with beautiful colour illustrations. Issued at irregular intervals.

Hawaiian Shell News.
Published by the Hawaiian Malacological Society. Obtainable from 2777 Kalakaua Avenue, Honolulu, Hawaii, 96815.
A well-illustrated newsy monthly magazine with many articles of interest to Australian collectors.

Journal of the Australian Malacological Society.
Edited by Mr Robert Burn and issued annually.
Obtainable from the Hon. Secretary, Australian Malacological Society, c/- The Australian Museum, P.O. Box A285, South Sydney, NSW. 2000.
This is a scientific journal containing original studies on Australian molluscs by amateur and professional malacologists.

The Veliger.
Edited by Dr Rudolph Stohler.
Published by the California Malacozoological Society Inc.
Obtainable from Mrs Jean Cate, 12719 San Vincente Boulevard, Los Angeles, California, 90049, USA.
Contains original research papers, some of which are useful to Australian collectors.

ear shells or abalones (FAMILY HALIOTIDAE)

THE ABALONES OR HALIOTIDS possess many characteristics considered to be "primitive" and the family is classified in a low position in the systematic arrangement of the gastropods. There are about 50 living species in the world and at least 23 of these occur in Australia.

As in most other gastropods, water is swept into the mantle cavity of abalones by ciliary action, passing through the gills where oxygen is absorbed and out again, carrying with it respiratory, excretory and alimentary wastes. In abalones the out-flowing water currents pass through the holes arranged in a row near the edge of the flat, coiled shell. These respiratory holes are a characteristic feature of abalones. Another is the beautiful nacreous (mother-of-pearl) lining on the inner surface.

Abalones are browsers, feeding on algae which they rasp from the rocks with their rather unspecialized radulae. They reproduce by mass population spawning when eggs and spawn are shed into the water. Fertilization takes place in the water and there follows a fully pelagic larval stage. Many species are gregarious and some of the larger gregarious species support commercial fisheries. Abalone meat is excellent eating and is canned for export or sale on the local market. The commercially exploited Australian species are *H. ruber* (N.S.W., Vic., Tas.), *H. roei* (southern W.A.) and *H. laevigata* (S.A., southern W.A.). The last-named species is not illustrated in this book.

1. *Haliotis ruber* LEACH, 1814
Ovate with a low spire, rounded dorsal surface. Sculptured with weak growth striae and broad but very irregular obliquely radiating folds crossed by numerous fine beaded spiral cords. Respiratory holes on conical tubercles, 6 or 7 holes open. There is a concave zone between respiratory holes and margin. Red-brown with narrow green curved radiating rays.
    16 cm. A common commercial species. N.S.W., Vic., Tas. and S.A.

2. *Haliotis conicopora* PÉRON, 1816
Ovate, spire of medium height, dorsal surface rather flat. Sculptured with irregular discontinuous obliquely radiating folds crossed by weak spiral striae, radial growth striae strong. Respiratory holes on high conical tubercles, 6 or 7 holes open. There is a concave zone between respiratory holes and margin. Green or red-brown with greenish patches and broad green curved radiating rays.
    20 cm. Uncommon. Vic. to Fremantle, W.A. *H. vixlirata* COTTON, 1943, and *H. granti* PRITCHARD and GATLIFF, 1902, are synonyms.

3. *Haliotis emmae* REEVE, 1846
Ovate, spire low. Sculptured with strong rough spiral riblets and irregular obliquely radiating folds which become weaker near respiratory holes, radial growth striae weak. Respiratory holes on conical tubercles, 6 or 7 holes open; below respiratory holes is a narrow spiral channel and several angular spiral ribs nearer the margin. Orange-red with curved cream radiating rays.
    10 cm. Common. Vic., Tas. and S.A.

4, 4a. *Haliotis scalaris* LEACH, 1814
Ovate, rather thin, spire moderately raised. Complexly sculptured with a broad, raised, central spiral rib which is itself sculptured with rough scaled spiral riblets and numerous thin, wall-like radial lamellae. A deep smooth concavity separates central rib from respiratory holes, and another deep smooth concavity separates respiratory holes from margin, margin rough and nodulose. Respiratory holes on very high conical tubercles, 4 to 6 holes open. Orange-red or orange-brown with cream or light fawn patches.
    10 cm. Moderately uncommon. S.A. to Geraldton, W.A.

5, 5a. *Haliotis elegans* PHILIPPI, 1899
Elongate-ovate, spire small, low, situated near posterior margin, dorsal surface rounded. Sculptured with heavy raised spiral ribs. Respiratory holes simple, 8 or 9 holes open in juveniles but in adults hole formation is irregular. Red-brown or orange with radiating rays of cream.
    10 cm. Uncommon. Cape Leeuwin to Shark Bay, W.A.

6, 6a. *Haliotis coccoradiata* REEVE, 1846
Ovate, rather thin, spire low, dorsal surface rather flat. Sculptured with numerous rough spiral cords crossed by very fine radial striae. Respiratory holes simple, 6 or 7 holes open. Red-brown with broad, irregular curved radiating cream rays.
    5 cm. Common. N.S.W. and Vic.

7, 7a. *Haliotis squamata* REEVE, 1846
Ovate to elongate-ovate, solid, spire small, low and situated near posterior margin, dorsal surface rounded. Sculptured with numerous beaded spiral riblets crossed by growth striae. Respiratory holes simple, 7 or 8 holes open. Dark red-brown with green spiral bands and zigzag spiral cream rays.
    6 cm. Common. Shark Bay to Broome, W.A.

8. *Haliotis cyclobates* PÉRON, 1816
Sub-circular, spire high, dorsal surface rounded. Sculptured with weak obliquely radiating folds and strong rough spiral riblets. Respiratory holes simple, 5 or 6 holes open. Below respiratory holes is a narrow shallow concavity and between that and the margin are oblique rows of beaded ribs. Brown and green with irregular curved radiating cream rays.
    6 cm. Common. Vic. to the Recherche Archipelago in southern W.A.

9, 9a. *Haliotis semiplicata* MENKE, 1843
Ovate to elongate-ovate, thin, spire low, dorsal surface rather flat. Sculptured with broad but low and widely spaced radiating folds, a low rounded spiral fold centrally, and fine spiral striae. Respiratory holes simple, 6 or 7 holes open. There is a smooth zone with angular spiral ribs near the margin below the respiratory holes. Green or fawn often with a broad orange spiral band centrally, and radial bands of pink and brown separated by red lines in zone near margin below the respiratory holes.
    6 cm. Uncommon. Esperance to Fremantle, southern W.A.

10, 10a. *Haliotis ovina* GMELIN, 1791
Ovate, rather thin, spire low, dorsal surface low but rounded. Sculptured with weak spiral striae and radiating, low, rounded folds which sometimes form spiral rows of nodules. Respiratory holes on conical tubercles giving spire a coronate appearance, about 4 holes open. There is a shallow concavity below the respiratory holes and several narrow ribs near the margin. Green or brown with broad curved radiating rays of yellow or white.
    6 cm. Common. Indo-West Pacific; eastern Qld.

11, 11a. *Haliotis varia* LINNAEUS, 1758
Ovate, rather solid, spire high, dorsal surface rounded. Sculptured with irregular radial folds crossed by spiral ribs of different thicknesses. Respiratory holes on low conical tubercles, 4 or 5 holes open. Usually brown or greenish with cream, white or green patches.
    4.5 cm. Common. Indo-West Pacific; Houtman Abrolhos, W.A. to eastern Qld.

12. *Haliotis roei* GRAY, 1826
Ovate, solid, spire usually low, but high in specimens living in situations exposed to heavy surf. Sculptured with thick, rough, spiral cords of different widths. Respiratory holes simple, about 7 open. A zone below the respiratory holes is rounded and sculptured with rough spiral ribs. Red-brown.
    12 cm. A common commercial species. Vic. to Shark Bay, W.A.

Plate 8. Haliotidae ($\frac{7}{10}$ × natural size)

top shells (FAMILY TROCHIDAE)

THIS VERY LARGE FAMILY comprises several subfamilies, many genera and hundreds of species. Many of the species are small, particularly those of the southern Australian shores. Only a few larger representatives of the family are illustrated here. The shells of trochids are usually conical, sometimes turbinate, lenticular or ovate, and have an inner nacreous (mother-of-pearl) layer. The operculum is round, multispiral and horny. Trochids are herbivorous molluscs, often gregarious. Most of them live among seaweeds or weed covered rocks. In some species fertilization is external, but in others the females lay the fertilized eggs in a soft gelatinous egg-mass within which the early stages of development take place.

1, 1a. *Angaria tyria* REEVE, 1843
Turbinate, deeply umbilicate, spire low with a flat apex, shoulders of whorls angulate. Width slightly greater than height. Entire external surface finely spirally ribbed, the ribs bear crowded crescent-shaped scales or prickles. Long curved spines present on the shoulders of the early whorls, but spines usually more-or-less obsolete on the body whorl of adult specimens. Off-white or grey except for a red or purple sutural band and purple umbilical area.
7 cm diameter. Common. Cockburn Sound to Shark Bay, W.A. The common northern Australian species *A. delphinus* LINNAEUS, 1758 is almost uniformly black and is much more heavily spined. Members of this genus are often mistaken for turban shells but the round horny operculum shows them to be trochids.

2, 2a. *Calliostoma ciliaris* MENKE, 1843
Thin, conical, with pointed apex, flat sides and base, and sharply angulate basal margin. Width equal to or slightly greater than height. Outer lip and columella smooth, aperture weakly lirate. Umbilicus sealed. Surface smooth, shiny, and sculptured with punctate striae, short and weak axial folds above and below the sutures, and a sharp marginal keel which forms a spiral rib above the sutures. Pink-fawn with patches of darker red-brown around the basal margin and sutures, and spiral lines of alternating brown and white dashes on the base and sometimes on the sides.
3.5 cm diameter. Uncommon. Southern W.A. from Esperance to Fremantle.

3, 3a. *Clanculus undatus* LAMARCK, 1816
Solid, widely umbilicate, conical, depressed, with low spire, rounded whorls and sub-angulate basal margin. Width greater than height. Roughly sculptured with beaded spiral ribs. Inner margin of outer lip irregularly toothed, aperture spirally lirate within. Columella with strong tooth-like folds. Brown with many dark red spots.
4 cm diameter. Moderately uncommon. N.S.W. to Cape Naturaliste, W.A. There are many smaller species of this genus in southern Australia.

4, 4a. *Thalotia chlorostoma* MENKE, 1843
Moderately thin, conical, with high spire, flat sides and base and sharply keeled basal margin. Basal keel of early growth stages remains evident as a spiral rib at the sutures of the spire whorls. Height slightly greater than width. Umbilicus sealed, columella smooth except for a projecting tooth-like corner near the base. Outer lip weakly toothed within the aperture. Colour variable, usually red or green with darker spots or arrows on the ribs and large dark maculations and pale stripes on the basal keel.
3 cm high. Moderately common. S.A. to southern W.A. as far north as Geraldton.

5, 5a. *Calliostoma australe* BRODERIP, 1835
Moderately thin, conical, with high spire, flat sides and base, and sharply keeled basal margin. Height greater than width. Umbilicus sealed, columella and outer lip smooth. Narrow, beaded spiral ribs adorn the whorls and base, and a thick spiral rib crossed by axial folds runs around the basal keel and above the sutures of the spire. Base red-brown, sides cream with broad axial rays of brown, beads on the ribs brown or mauve.
3 cm high. Uncommon. Vic. to Tas. to Fremantle, W.A.

6, 6a. *Calliostoma monile* REEVE, 1863
Moderately solid, conical, with high spire, flat sides and base, and weakly keeled sub-angulate or rounded basal margin. Height greater than width. Sculptured with fine spiral striae and a weak rib at sutures of spire whorls. Columella and outer lip smooth. Cream with mauve spots on the spiral rib at the sutures and around the base.
2.5 cm high. Moderately common. Monte Bello Is., W.A. to eastern Qld.

7, 7a. *Austrocochlea rudis* GRAY, 1826
Solid and turbinate, spire high, whorls rounded. Umbilicus closed. Height and width about equal. Aperture spirally lirate internally and denticulate near the margin. Exterior smooth. Dark purple-grey, lip black or dark green.
3.5 cm high. Abundant. Intertidal, rocky shores. S.A. to Murchison River, W.A.

8, 8a, 8b, 8c. *Austrocochlea constricta* LAMARCK, 1822
Solid and turbinate, with moderately high spire, and rounded whorls and basal margin. Umbilicus sealed. Outer lip and columella smooth except for a weak knob-like tooth which is sometimes present on anterior part of the columella. Body whorl encircled by 5 or 6 strong spiral ribs. Colour variable, may be uniform purple-grey (8, 8a) or grey with wavy axial bands of dark purple-brown (8b, 8c).
2.5 cm high. Common. N.S.W. to Houtman Abrolhos, W.A. *A. concamerata* WOOD, 1828 is like this and often found in the same places, but may be distinguished by its flatter shell, stronger ribbing and yellow spotted or streaked exterior.

9, 9a. *Monodonta labio* LINNAEUS, 1758
Solid, turbinate, with rounded whorls and basal margin and moderately high spire. Umbilicus sealed. Outer lip toothed along its inner margin, columella bears a large squared tooth. Whorls and base sculptured with strong granulated spiral ribs. Grey, green, brown or red with darker spots on the ribs.
3.5 cm high. Common Indo-West Pacific; Dampier Archipelago, W.A. to eastern Qld.

10, 10a, 10b, 10c. *Phasianotrochus eximius* PERRY, 1811
Moderately solid, elongately turbinate, with a high pointed spire, and rounded whorls and basal margin. Height greater than width. Surface glossy but encircled by moderately strong striae, between which are very faint striae. Columella smooth, nearly vertical. Outer lip smooth, aperture strongly lirate, lirae sometimes end in smooth pointed nodules near lip. Colour variable, usually fawn, olive-green, brown or rose, sometimes patterned with pale lines or stripes, interior brightly iridescent.
4 cm high. Common. N.S.W. to Fremantle, W.A.

11, 11a. *Trochus (Trochus) maculatus* LINNAEUS, 1758
Solid, umbilicate and conical, with almost flat sides and base although the basal margin may be rounded or only sub-angulate. Height and width approximately equal. Columella thickened, with about 5 strong spiral folds which end in broad rounded tubercles at the margin. Outer lip smooth, aperture finely spirally lirate with strong spiral ridges near the base. Sides roughly sculptured with beaded spiral ribs and axial striae, while the base has finer, more regular spiral ribs and many fine obliquely radiating striae. Cream to white with red-brown or green axial flames.
6 cm high. Indo-West Pacific; Houtman Abrolhos, W.A. to eastern Qld.

12, 12a. *Trochus (Tectus) pyramis* BORN, 1778
Solid, conical, with flat sides and base, sharply angulate basal margin and indented sutures. Umbilicus sealed, columella smooth but curved and ends in a thick, folded, spirally curved tubercle. Outer lip smooth except for several prominent denticles on inner margin near the base. Height and width approximately equal. Sides usually smooth except for growth lines crossed by fine oblique-axial striae and weak nodules at sutures, but some specimens weakly spirally ribbed. Base with fine spiral striae. Sides grey, cream, pink or green. Base white, blue-green or green.
8 cm high. Common. Indo-West Pacific; Rottnest Is., W.A. to eastern Qld. *Trochus obeliscus* GMELIN, 1788 is a synonym.

13, 13a. *Trochus (Trochus) lineatus* LAMARCK, 1822
Solid, umbilicate, conical with flat sides and base and moderately or sharply angulate basal margin. Height and width approximately equal. Columella smooth or weakly toothed. Sides sculptured with spiral ribs bearing regularly spaced, obliquely enlongated fine nodules, and a double spiral rib with larger nodules at the sutures. Base finely sculptured with spiral cords and oblique striae. Cream or green with red oblique-axial lines.
5 cm high. Common. Coral and rocky reefs in shallow water. Indo-West Pacific; Pt. Cloates, W.A. to eastern Qld.

Plate 9. Trochidae (natural size)

turban and star shells (FAMILY TURBINIDAE)

THE SHELLS OF THIS FAMILY are turbinate or conical and have an inner nacreous layer like that of their close relatives the trochids. In many other respects these 2 families are similar. The main distinguishing character is the presence of a solid, heavy calcareous operculum in the Turbinidae, while in the Trochidae the operculum is horny. Sculpture and shape of the operculum are often useful indentification characters. The columella and outer lip of turban shells are usually smooth. Members of the Turbinidae are herbivorous. The females lay gelatinous egg-masses from which the young hatch as early stage free swimming larvae. Most species live in shallow water among seaweeds and algae, or on coral reefs.

1. *Turbo (Marmarostoma) chrysostomus* LINNAEUS, 1758
Solid and turbinate. Spire moderately high, whorls rounded, axially squamose and spirally ribbed. Ribs on the body whorl sometimes spinose. Umbilicus sealed or nearly so. Exterior light brown with flames and patches of cream, darker brown and green, inner edge of lip orange-yellow, columella and deep interior white. Operculum thick and circular, its convex outer surface smooth and brown, yellow or green at the centre, but obliquely striate and paler toward the margin.
6 cm high. Common. Indo-West Pacific; eastern Qld.

2, 2a, 2b, 2c. *Turbo (Marmarostoma)* cf. *argyrostomus* LINNAEUS, 1758
Solid and turbinate with moderately high spire, impressed sutures and rounded, axially squamose and spirally ribbed whorls. The spiral rib at the shoulders, and 1 or 2 of those at the centre of the body whorl prominent and usually bear short hollow spines. A thick spinose funicle surrounds the deep umbilicus. Operculum thick, circular, and granulose at the centre with short, fine, oblique folds around the margin. Exterior fawn or brown with darker patches and, usually, bright green ribs and spines. Inner edge of the lip green, but the deep interior, columella and outer surface of the operculum white.
10 cm. Common. Pt. Cloates to Barrow Is., W.A. The specimens illustrated are from Pt. Cloates. They are more heavily ribbed than the typical *T. argyrostomus* of Qld. and other parts of the Indo-West Pacific, but a close affinity with that species seems certain. Spine development varies as indicated by the difference between (2) and (2c) specimens from the same locality, but when present the spines are strong and widely spaced, and unlike the fine prickles of typical *T. argyrostomus*.

3. *Turbo (Marmarostoma) pulcher* REEVE, 1842
Solid, turbinate, with a moderately high spire, rounded whorls, and sealed umbilicus. Height greater than width. Sculptured with spiral ribs crossed by closely packed, thin, scale-like axial lamellae. Operculum thick, circular, with blunt club-shaped tubercles covering the convex outer surface. Exterior fawn with wavy brown axial stripes and sometimes green or orange mottling, aperture, columella and outer surface of operculum white.
8 cm height. Common. Southern W.A. from Esperance to Shark Bay. A distinctive cool-water member of an otherwise tropical genus.

4, 4a. *Subninella undulata* SOLANDER, 1786
Solid, deeply umbilicate, turbinate with moderately low spire and rounded whorls. Width greater than height. Exterior has strong, smooth, spiral ribs, green with wavy oblique-axial darker green rays. Operculum thick, circular, with a smooth, domed outer surface.
7 cm diameter. Common. N.S.W. to Hopetoun, southern W.A.

5, 5a, 5b. *Ninella torquata* GMELIN, 1791
Solid, deeply umbilicate, turbinate, with moderately low spire and deep sutures. Early whorls keeled, the keel persisting on the body whorl of most western specimens (5b) but becoming obsolete on the body whorl in eastern specimens (5). Width greater than height. Sculptured with closely packed, thin, high, axial lamellae. Operculum (5a) thick and circular, the outer side bearing fine prickles and raised spiral ridges. Exterior sand-coloured, sometimes with green or orange mottling in young specimens, interior, columella and outer side of operculum white.
11 cm diameter. Common. N.S.W. to Geraldton, W.A. The name *N. whitleyi* IREDALE, 1949, has been applied to heavily keeled specimens from W.A. (5b) but intermediate forms occur along the south coast.

6, 6a. *Turbo (Dinassovica) jourdani* KIENER, 1839
A very large, solid, turbinate shell, with sealed umbilicus, moderately high spire, impressed sutures and rounded whorls. Height slightly greater than width. Surface smooth except for irregular growth lines and weak spiral ribs on the early whorls. Operculum (6a) circular, thick and heavy, with a smooth porcellaneous-white outer surface. Early whorls cream with brown patches and interrupted brown spiral lines, penultimate and body whorls rich red-brown.
20 cm high. Moderately uncommon. S.A. to Geraldton, W.A. The largest of the Australian turban shells. The illustrated specimen is a juvenile.

7, 7a. *Turbo petholatus* LINNAEUS, 1758
Solid, turbinate, with rounded whorls, moderately high spire and sealed umbilicus. Height equal to or slightly greater than width. Operculum thick, rounded and smooth. Shell surface smooth and highly polished. Colour variable, usually brown or green with darker wavy axial zones and broad dark spiral lines which contain pale arrowhead marks or axial bars. Columella yellow, orange or green, outer surface of operculum with a green centre and brown margin.
6 cm height. Common. Indo-West Pacific; North-West Cape, W.A. to eastern Qld. The operculum is known as the "cat's eye" and is used extensively in shell jewellery.

8, 8a. *Astraea pileola* REEVE, 1842
Solid, depressed, conical, with a sealed umbilicus. Sides flat, except for an expanded, thin, fluted flange which runs around the basal margin and above the sutures. Width double the height. Base sculptured with scaled spiral ribs but the sides of the whorls are smooth. Exterior fawn to grey, interior pearly-white. Operculum oval, with a smooth outer surface, a very dark purple-brown centre and a lighter green rim.
6.5 cm diameter. Common. Exmouth Gulf to Broome, W.A.

9, 9a. *Astraea rotularia* LAMARCK, 1822
Solid, depressed, and conical, with sealed umbilicus and keeled basal margin. Width nearly double height. Whorls sculptured with oblique ridges and prominent, imbricated, laterally compressed nodules around the basal margin and sutures. Base sculptured with scaled spiral ribs. Exterior fawn or grey, interior white. Operculum oval, smooth, dark green to purple-brown.
4.5 cm diameter. Common. Port Hedland, W.A. to N.T.

10. *Phasianella ventricosa* SWAINSON, 1822
Moderately thin, turbinate, with sealed umbilicus, moderately high spire, impressed sutures and highly polished, smooth and rounded whorls. Height about 1½ times the width. Operculum white, smooth and ovate, pointed at one end. Several distinct colour patterns occur: (a) the most common, rose-pink with irregular wavy brown axial bands, flecks of yellow and numerous thin spiral lines of brown and yellow dashes; (b) like (a) but the brown axial bands broad and more regular, giving the shell an axially striped appearance; (c) red-brown background with splashes of pink or yellow and thin spiral lines of alternating long brown dashes and yellow spots; (d) uniform red-brown; (e) uniform rose-pink with faint spiral interrupted lines and a broad central band of red-brown blotches.
4 cm high. Common. N.S.W. to about Geraldton, W.A. This species is like *P. australis* but stouter. *P. perdix* WOOD, 1828 is a synonym.

11, 11a, 11b. *Phasianella australis* GMELIN, 1791
Moderately thin, elongately turbinate, with sealed umbilicus, high pointed spire, impressed sutures, and highly polished, smooth, rounded whorls. Height almost twice the width. Operculum white, smooth, ovate and pointed at one end. Several distinct colour forms occur: (a) the most common, clouded rose or fawn with spiral bands containing red blotches, axial stripes and arrow-like marks (11b); (b) clouded rose or fawn background with broad oblique-axial bands of red-brown, bands bounded by wavy white lines and crossed by narrow spiral white or pink lines; (c) like (b) but the oblique-axial bands narrower and broken into rows of rectangular blotches by prominent spiral lines, giving a distinct reticulate appearance (11); (d) uniform yellow with spiral rows of closely packed, fine, short, red axial lines (11a); (e) pale yellow or white shells without markings.
10 cm high. Common. Vic. and Tas. to Geraldton, W.A.

Plate 10. Turbinidae (⁸/₁₀ × natural size)

1

2

2a

2b

2c

3

4

4a

5

5a

6

6a

7

7a

8

8a

9

9a

10

11

11a

11b

5b

nerites (FAMILY NERITIDAE)

MEMBERS OF THIS FAMILY occupy an amazing variety of habitats. The majority of species are gregarious intertidal creatures of the marine rocky shores, but some species live in estuaries and some even in freshwater streams and lakes. Except for a few limpet-like freshwater species, nerites have thick globular shells with low spires and semi-circular apertures which are usually strongly toothed. The operculum is thick and calcareous and is locked in place by a peg-like tooth on the columellar side (6a). There is usually a broad flat shelf or "deck" adjacent to the columella on the underside of the body whorl. This may be smooth, pustulose or lirate, characters which are useful for identification purposes. Nerites are herbivorous molluscs feeding on algae or algal slime growing on rocks or logs. The females of marine species lay dome-shaped egg-masses which they attach to the substrate or to the shells of their comrades. In *N. atramentosa, N. albicilla* and many other nerites there is a planktonic larval stage but in some species development is direct (Anderson, 1962, Proc. Linn. Soc. New South Wales, 87: 62-68).

1, 1a. *Nerita polita* LINNAEUS, 1758
Spire flat, diameter greater than height. Surface shiny but sculptured with thick growth lines. Columellar deck smooth, but several weak, rather rounded teeth present at the centre of the columellar margin. Operculum shiny smooth except for a finely radially striate zone along the margin. Exterior cream, brown, pink or green, with irregular black or brown markings, aperture orange, columellar deck red or dark orange, operculum black, grey or fawn.
3 cm diameter. Abundant. Indo-West Pacific; Broome, W.A. to northern N.S.W.

2, 2a. *Nerita atramentosa* REEVE, 1855
Wide, with flat spire and dull, sometimes finely striated surface. Outer lip finely serrated along its inner margin, with a rounded nodule near the posterior end. Columellar deck narrow, smooth or with a few weak lirae, and 2 to 4 weak teeth on the columellar margin. Exterior black, aperture and columellar deck white, operculum black or grey.
3 cm diameter. Abundant. Southern Qld. to Pt. Cloates, W.A. This is the only nerite in southern Australia. *N. melanotragus* SMITH, 1884 is a widely used synonym.

3, 3a. *Nerita albicilla* LINNAEUS, 1758
Spire flat, shell almost planispiral. Diameter considerably greater than height. Surface dull, with strong spiral ribs and fine wavy growth lines. Columellar deck calloused and pustulose, operculum finely granulose. Exterior black, or grey and white with black patches, columellar deck yellow or white, operculum grey or yellow.
2.5 cm diameter. Common. Indo-West Pacific; Shark Bay, W.A. to northern N.S.W.

4, 4a, 4b. *Nerita undata* LINNAEUS, 1758
Turbinate with moderately high spire. Height greater than diameter. Exterior with numerous fine spiral riblets. Many fine teeth present on inner margin of outer lip and 1 or 2 moderately strong teeth posteriorly. Inner margin of the columella bears 3 or more strong central teeth, columellar deck strongly wrinkled. Operculum finely pustulose. Exterior usually yellow, light brown, grey or black with green, grey or black spots or patches, aperture white, often stained yellow, especially on the columellar deck, operculum grey.
4 cm high. Abundant. Indo-West Pacific; Geraldton, W.A. to eastern Qld.

5, 5a. *Nerita plicata* LINNAEUS, 1758
Rather high-spired, turbinate, and sculptured with coarse spiral ribs. Outer lip thick and strongly toothed. Columellar deck calloused, wrinkled and rounded, with 4 strong, squared teeth on the columellar margin. Operculum smooth. Exterior white or yellow, sometimes spotted with grey, outer lip and columellar deck white, deep interior yellow, operculum fawn.
3 cm high. Common. Indo-West Pacific; Houtman Abrolhos, W.A. to northern N.S.W.

6, 6a, 6b. *Nerita costata* GMELIN, 1791
A high-spired turbinate shell sculptured with 12 to 15 broad, rough spiral ribs. Edge of outer lip sharp and fluted, inner side strongly toothed with one particularly large tooth at the posterior end. Columellar deck transversely lirate posteriorly, strongly pustulose anteriorly, with 4 or 5 large teeth on the columellar margin. Operculum minutely granulose. Exterior dull black, sometimes with yellow lines between the ribs, interior white, operculum grey (6a).
3.5 cm high. Common. Indo-West Pacific; eastern Qld.

7, 7a. *Nerita lineata* GMELIN, 1791
Wide, low-spired, sculptured with numerous fine spiral ribs. Aperture sharp-edged and weakly toothed within. Columellar deck smooth but with several weak teeth centrally on the inner margin. Operculum finely granulose. Exterior light brown with darker spiral lines, columella yellow or orange, operculum grey.
4 cm diameter. Common. Indo-West Pacific; Shark Bay, W.A. to eastern Qld.

8, 8a, 8b, 8c, 8d, 8e, 8f, 8g. *Nerita chamaeleon* LINNAEUS, 1758
Broad, low-spired, roughly sculptured with strong rugose spiral ribs, sometimes with finer ribs between them. Aperture sharp-edged and weakly toothed within. Columellar deck narrow, with a few weak pustules and lirae and 2 to 4 small central teeth on inner margin. Operculum finely granulose. Exterior uniformly yellow, red, orange, grey, black or white, often banded, maculated, or spotted with grey, black or purple. Interior and columellar deck white, operculum grey.
2 cm diameter. Abundant. Indo-West Pacific; North-West Cape, W.A. to eastern Qld. The extreme colour variation of this species is illustrated here by a selection from a single locality (Woody Is., Qld.).

periwinkles (FAMILY LITTORINIDAE)

IN EUROPE AND NORTH AMERICA periwinkles are a popular sea food. Most Australian species are too small to be of much edible value. The family is represented in tropical, temperate and cold regions in all parts of the world. As the name implies, littorinids are inhabitants of the littoral (intertidal) zone, although some species live well above high tide level and are only occasionally splashed by waves. They browse on seaweeds or on microscopic algae growing on rocks, logs or mangrove trunks.

9. *Tectarius pagodus* LINNAEUS, 1758
Conical, whorls encircled by fine spiral striae and heavily nodulose or tuberculate spiral ribs. The rib around the basal margin and above the sutures bears large laterally flattened tubercles which form a keel. Outer wall of aperture spirally lirate. Exterior yellowish, interior pale yellow or pink, columella white.
4 cm high. Common. Central Indo-West Pacific; Broome, W.A. and N.T.

10, 10a. *Nodilittorina rugosa* MENKE, 1843
Turbinate to ovate with a moderately high spire, rounded whorls, sutures impressed and a rough exterior, sculptured with spiral striae, and often low axial nodulose folds. Exterior yellow-grey, sometimes bluish; interior and columella yellow or light brown.
2 cm high. Abundant. Esperance to Shark Bay, W.A.

11, 11a. *Nodilittorina pyramidalis* QUOY & GAIMARD, 1833
Turbinate to conical. Spire moderately high, whorls rounded, and sculptured with fine spiral striae and 2 spiral rows of rather pointed nodules. Exterior blue-grey with fawn coloured nodules, interior and columella brown.
1 cm high. Abundant. Western, northern and eastern Australia from Fremantle to eastern Vic.

12, 12a. *Melaraphe unifasciata* GRAY, 1826
Elongate-turbinate with a high spire, rounded whorls, and moderately impressed sutures. Sculptured with very fine spiral striae. Exterior white with a broad central pale blue spiral band, interior and columella brown.
2.5 cm high. Abundant. Southern Australia from southern Qld. to North-West Cape, W.A.

13, 13a. *Bembicium auratum* QUOY & GAIMARD, 1834
Wide, depressed, conical and flat based, with a sharply angulate basal margin, and rugose spiral ribs on the whorls. Basal margin and sutural ridge tuberculate. Exterior pale yellow or green-brown with grey mottling or wavy grey axial-oblique stripes, base pale yellow-brown, columella golden brown, interior chocolate brown.
2 cm diameter. Abundant. Southern Australia from southern Qld. to Houtman Abrolhos, W.A.

Plate 11. Neritidae and Littorinidae (1²/₁₀× natural size)

ceriths (FAMILY CERITHIIDAE)

CERITHS, SOMETIMES ALSO CALLED CREEPERS, generally have long, tapered many-whorled shells, often sculptured and sometimes very colourful. Characteristically, they have a moderately long up-turned anterior canal and a horny operculum which is sub-circular and has an off-centre nucleus. Most ceriths live gregariously on sand or weed, or among coral rubble in shallow water. They are herbivorous or detrital-feeding creatures. The females lay their eggs in twisted strings of soft jelly from which the young hatch as free-swimming larvae. The family is most strongly represented in tropical waters where most of the largest species also occur, but there are several small species and one giant in southern Australia. A few larger ceriths, representative of the family, are illustrated here.

1. *Campanile symbolicum* IREDALE, 1917
Very large, elongate, acutely turreted and with slightly concave sides. Whorls smooth except for sinuous growth-lines and a low spiral ridge just behind the sutures. Anterior canal short, nearly horizontal. Columella smooth and sinuous, outer lip smooth. Exterior chalky-white, interior glossy white.
20 cm. Common. Esperance to Geraldton, W.A. The genus has a long fossil record but this is the only surviving species.

2, 2a. *Cerithium (Aluco) cumingi* A. ADAMS, 1855
Elongate, with a deep posterior channel, and a long and nearly vertical anterior canal. Early whorls spirally ribbed and nodulose, later whorls and body whorl smooth. Columella calloused. Exterior white or cream, profusely spotted and axially striped with dark red-brown, interior and columella white.
10 cm. Common. Indo-West Pacific; North-West Cape, W.A. to eastern Qld. (?). *C. aluco* LINNAEUS, 1758 is similar but shorter and all the whorls are nodulose.

3, 3a, 3b, 3c, 3d, 3e. *Cerithium fasciatum* BRUGUIÈRE, 1792
Solid, slender and acutely tapered with a high, nearly vertical anterior canal. Early whorls faintly axially lirate, later whorls glossy, smooth, sculptured only with fine spiral striae. Columella calloused, with a broad low spiral fold at its centre. External colour extremely variable, usually white with brown spiral lines or bands, sometimes uniformly white, interior and columella white.
9 cm. Abundant. Indo-West Pacific; Pt. Cloates, W.A. to eastern Qld.

4, 4a, 4b. *Rhinoclavis vertagus* LINNAEUS, 1758
Rather stout but with a sharp, pointed spire and slightly convex sides. Whorls smooth anteriorly, but with strong short axial folds near sutures. Anterior canal long, up-turned, and almost vertical. Columella thickly calloused, with a strong spiral fold at its centre and a weak parietal ridge near the narrow posterior channel. Exterior white or cream, often stained brown, interior and columella white.
8 cm. Abundant. Indo-West Pacific; Houtman Abrolhos, W.A. to eastern Qld.

5, 5a. *Cerithium nodulosum* BRUGUIÈRE, 1792
Elongate and heavily sculptured, each whorl bearing a single spiral row of prominent tubercles and a number of weakly nodulose spiral ribs. Strong nodulose spiral ribs present at the anterior end of body whorl. Anterior canal short, oblique. Outer wall of aperture spirally channelled and there is a strong parietal ridge beside a deep posterior channel. Exterior white with brown splotches but usually thickly covered with calcareous growths, interior and columella white.
1.5 cm. Abundant. Indo-West Pacific; N.T. to eastern Qld.

6, 6a. *Rhinoclavis bituberculatum* SOWERBY, 1865
Rather stout, sharp spired. Whorls finely spirally striate and bear several weakly nodulose spiral ribs. Anterior canal short but almost vertical. Columella heavily calloused, with a strong central spiral fold and a low parietal ridge beside the posterior channel. Exterior fawn, spiral ribs and nodules white, apex blue-grey, interior and columella white.
6 cm. Common. Cape Leeuwin to Broome, W.A. The spiral ribs are more heavily nodulose in specimens from the north of W.A. (6) than in southern specimens (6a).

mudwhelks (FAMILY POTAMIDIDAE)

THE MUDWHELKS, SOMETIMES ALSO CALLED CREEPERS, are close relatives of the ceriths which they resemble both in shell form and habitat. Distinction between the 2 families is based mainly on anatomical and radula characters, although there are some differences in the shells. Mudwhelks usually have flared outer lips, a more complex columella structure, and a very short anterior canal which is not up-turned as it is in the ceriths. The operculum is horny with a central nucleus. Whereas most ceriths live in sand, most species of the Potamididae live on muddy shores near high tide level and are capable of living for long periods out of water. Like the ceriths, mudwhelks are gregarious herbivores or detrital feeders. The females lay gelatinous egg-strings and there is a planktonic larval stage.

7, 7a. *Telescopium telescopium* LINNAEUS, 1758
Conical, base broad and rather flat, sides straight. Whorls short and sculptured with several deep spiral striae. Anterior canal and columella very short. Columella twisted, with a strong central spiral ridge. Outer lip thin, not flared, smooth except for weak serrations anteriorly. Exterior brown, often with a single spiral central white line on each whorl, columella yellow, interior brown.
11 cm. Abundant. Indo-West Pacific; Port Samson, W.A. to eastern Qld.

8, 8a. *Terebralia sulcata* BRUGUIÈRE, 1792
Whorls spirally grooved and bear broad, rounded, axially elongate ridges. Outer lip smooth, widely flaring, and completely surrounding the anterior canal so that the latter appears as a round hole in the anterior end of the shell. Columella weakly lirate. Exterior grey or brown, often stained green, lip white, deep interior rich brown.
7 cm. Abundant. Indo-West Pacific; Shark Bay, W.A. to eastern Qld.

9, 9a. *Velacumantus australis* QUOY & GAIMARD, 1834
Elongate, with deeply impressed sutures and a flared, toothed outer lip. Whorls sculptured with broad, axially elongate ridges crossed by numerous rough, nodulose spiral ribs. Anterior canal very short, columella smooth, interior weakly lirate. Exterior uniformly dark brown or grey, or with a central white spiral line, interior brown.
4 cm. Abundant. Southern Australia from southern Qld. to Fremantle, W.A. These animals are known to be secondary hosts for the parasitic flat-worm *Austrobilharzia*. The parasites' free-swimming larvae emerge from the snail host in vast numbers in summer and normally infect aquatic birds, but they sometimes burrow through the skin of humans causing lesions known as "schistosome dermatitis" or "swimmers' itch". However, this parasite cannot survive in a human host and the irritation it causes is brief.

10, 10a. *Pyrazus ebeninus* BRUGUIÈRE, 1792
Elongate, with finely, spirally ribbed whorls. Early whorls bear axially elongate nodules, later spire whorls and body whorl angulate and usually bear a single spiral row of high nodules or tubercles. Outer lip widely flared and weakly lirate internally. Anterior canal very short, posterior channel present. Columella smooth, calloused, notched on the left side. Exterior grey or brown, columella white, deep interior purple or brown.
11 cm. Abundant. Southern Qld. to Vic.

Plate 12. Cerithiidae and Potamididae ($\frac{8}{10}$ × natural size) ⇩

wentletraps (FAMILY EPITONIIDAE)

THE FAMILY EPITONIIDAE (sometimes known as the Scalidae) is represented in all tropical, sub-tropical and temperate seas. There are hundreds of species and it is difficult to tell some of them apart. Australian shores are particularly rich in wentletraps but it has been possible to illustrate only a few larger species here. The shells have many whorls, usually with strong axial varices. Some species also have strong spiral sculpturing. The aperture is usually round, sometimes with a raised rim, the umbilicus may be completely sealed or very deep. Often the whorls are very loosely coiled and are united only by narrow lamellae. There is a thin, horny, multi-spiral operculum. It has been shown that many wentletraps lead commensal lives in association with sea anemones or other coelenterates. They have a long proboscis and bite pieces from the coelenterate host.

1, 1a. *Granuliscala granosa* QUOY & GAIMARD, 1834
Elongate, high-spired, with many whorls, width more than ⅓ but less than ½ the height. Sutures impressed, umbilicus sealed. Varices low, rounded, about 10 per whorl. On body whorl, varices end at a weak spiral rib which runs around base. White.
    4 cm. Common. Vic. and Tas. to Fremantle, W.A. Do not confuse this with *O. australis* which is similar but has more prominent, raised varices.

2, 2a. *Opalia australis* LAMARCK, 1822
Elongate, high-spired, with many united whorls, impressed sutures and sealed umbilicus. Width about ⅓ height. About 8 strong varices per whorl, ending at a thick spiral rib around base of body whorl. Extremely fine spiral striae between the varices. White.
    4 cm. Common. N.S.W. to Fremantle, W.A.

3, 3a. *Epitonium perplexa* PEASE, 1860
Moderately thin. Height more than double the width. Sutures moderately deep, umbilicus closed. Varices low but strong, widely spaced, about 13 per whorl. White or fawn between varices, varices white.
    4 cm. Uncommon. Indo-West Pacific; Rottnest Is., W.A. (Qld.?).

4, 4a, 4b. *Epitonium pallasi* KIENER, 1838
Moderately stout, short, width about ⅔ height. Umbilicus very deep, sutures so deep that whorls are separated although the corresponding varices are joined. Varices strong, widely spaced, about 10 per whorl. Fawn or pale brown between the varices, varices white.
    2.5 cm. Moderately common. Indo-West Pacific; eastern Qld. *E. neglecta* ADAMS & REEVE, 1850, also of Qld., is very like this but the posterior ends of the varices are hooked and the varices are usually more numerous.

5, 5a. *Epitonium imperialis* SOWERBY, 1844 (see Fig. 2, page 9).
Thin and fragile. Width about ⅔ height. Sutures and umbilicus deep. Varices thin, fine, numerous, about 30 per whorl. Exterior pale fawn or purple-brown, varices white.
    4 cm. Moderately common. Indo-West Pacific; Cape Naturaliste, W.A. to southern Qld. Field observations near Fremantle suggest that the shells may be sexually dimorphic, the females being larger than males.

6, 6a. *Epitonium scalare* LINNAEUS, 1758
Moderately stout, width about ⅔ height. Umbilicus very deep, sutures so deep that whorls are separated although the corresponding varices are joined. Varices strong, widely spaced, about 8 per whorl. White or pale grey-fawn between varices, varices white.
    5 cm. Moderately common. Indo-West Pacific; eastern Qld. Once considered rare, it commanded high prices.

7, 7a. *Cirsotrema varicosa* LAMARCK, 1822
Slender and high spired, width about ⅓ height. Whorls rounded, sutures deep, crossed by thick lamellae. Umbilicus closed. Two very strong nodulose varices per whorl, plus high, flattened axial ribs connected by spiral ribs, with deep pits in the interspaces. White.
    4 cm. Uncommon. Indo-West Pacific; Houtman Abrolhos, W.A. to eastern Qld.

8, 8a. *Cirsotrema kieneri* TAPPARONE-CANEFRI, 1876
Very high spired and slender, width less than ¼ height. Sutures deep and wide, umbilicus closed. Fine and numerous axial ribs crossed by fine spiral ridges of almost equal size, so forming a cancellate pattern. White.
    8 cm. Uncommon. Indo-West Pacific; Dampier Archipelago, W.A. to eastern Qld.

sundials (FAMILY ARCHITECTONICIDAE)

THIS IS A SMALL FAMILY with species living in tropical and temperate seas. The shells are discoidal or depressed—conical with a rather flat base. Most of them are sculptured with beaded spiral and radial ribs and have a very widely open umbilicus within which can be seen the spiral coiling of the whorls. In the 2 genera *Architectonica* and *Philippia* the operculum is flat, horny, spirally coiled and has a tubercular knob on its inner surface. In *Torinia* (= *Heliacus*) the operculum is conical, chitinous and spirally coiled. Because of this difference, and differences in the radula, *Torinia* and its subgenera are sometimes placed in a separate family, the Heliacidae. *Grandeliacus* has an operculum like that of *Architectonica* although IREDALE (1957) allied the genus with *Torinia*.

9, 9a. *Architectonica perdix* HINDS, 1844
Conical, with rather tumid whorls, an angulate and ribbed periphery, and incised sutures. Sculpture like that of *A. perspectiva* except for less pronounced radial ribs. Marginal crenulations around the umbilicus white. Generally flesh coloured, with spiral rows of brown spots above and below sutures and peripheral keel.
    3.5 cm diameter. Common. Indo-West Pacific. Northern W.A. to eastern Qld.

10, 10a. *Architectonica reevei* HANLEY, 1862
Rather tumid and conical like *A. perdix* but has fine incised spiral lines between the main sutural and peripheral grooves, and a few brown spots on the umbilical teeth.
    2.5 cm diameter. Common. N.S.W. The name *A. offlexa* IREDALE, 1931 is a synonym.

11, 11a, 11b. *Architectonica perspectiva* LINNAEUS, 1758
Low, conical, periphery angulate and ribbed, sutures deeply incised. A single spiral groove encircles whorls just below sutures, spiral ribs (often beaded) present above and below the peripheral rib and around the wide umbilicus. Early whorls with radial ribs which disappear on later whorls. Margin of umbilicus bears heavy brown teeth or crenulations. Exterior fawn or ash-grey with spiral bands of white and brown, the brown bands often interrupted.
    5 cm diameter. Common. Indo-West Pacific; North-West Cape, W.A. to N.S.W. The N.S.W. form was named *A. perspectiva fressa* IREDALE, 1936.

12, 12a. *Architectonica maxima* PHILIPPI, 1848
Like *A. perspectiva* but heavier, with much stronger radial sculpture and a deep spiral groove near the centre of the whorls. Brown spots between the suture and the main sub-sutural groove but no brown spiral band, umbilical teeth white.
    5 cm diameter. Common. Indo-West Pacific; Qld. and N.S.W.

13, 13a. *Philippia radiata* RÖDING, 1798
Low, conical. Periphery weakly ribbed. Whorls smooth except for fine spiral striae above and below the peripheral rib. Umbilicus deep but narrow, bordered by low irregular crenulations. Upper surface white with a single spiral brown band below the suture, radial brown lines cross the whorls from the spiral band to the periphery. Base fawn or brown, umbilical area white.
    2 cm diameter. Common. Indo-West Pacific; Pt. Cloates, W.A. to eastern Qld.

14, 14a. *Grandeliacus moretensenae* IREDALE, 1957
Sub-discoidal with rounded whorls sculptured with fine radial grooves and about 16 strong spiral ribs. Umbilicus wide and deep and bordered by an angular weakly toothed rib. Uniformly brown.
    3 cm diameter. Common. Port Curtis to Moreton Bay, Qld.

15, 15a, 15b. *Torinia variegata* GMELIN, 1791
Depressed, discoidal to conical, with a deep but narrow umbilicus. Sculptured with strong nodulose spiral ribs, 2 particularly strong ribs form a keel around periphery. White or pale grey with large, conspicuous black or dark brown spots on ribs on upper surface.
    1.5 cm diameter. Common. Lives among zooanthids in shallow water. Indo-West Pacific; Rottnest Is., W.A. to eastern Qld.

Plate 13. Epitoniidae and Architectonicidae (⁹⁄₁₀ × natural size)

strombs, scorpion shells and little auger shells

(FAMILY STROMBIDAE)

MOST OF THE SEVERAL GENERA in this colourful family are represented on Australian shores. The largest genus is *Strombus* with about 50 species which are usually found in shallow water near coral reefs. There are some strombs in the tropical Atlantic, tropical eastern Pacific and in the Mediterranean, but the largest number of species occurs in the tropical Indo-West Pacific region. Northern Australia has at least 18 species, most of which are widespread in the Indo-West Pacific. Although there are no strombs peculiar to the cool waters of southern Australia, one species, *Strombus mutabilis*, is found far south of the tropics on both sides of the continent.

The operculum of strombs is long, sharp, sickle-shaped and usually serrated on one side. It is located on a posterior protuberance of the muscular foot and is used by the animal as a lever to push itself along or to kick itself over when it has been rolled on to its back. Sometimes it also uses the operculum as a weapon of defence.

Shells of adult strombs have a flaring or thickened outer lip which has a deep U-shaped notch near the anterior end. The notch is used as a peep-hole for the right eye. The large eyes of strombs are located on short branches near the end of usually long eye stalks (plates 15 and 16). The columella is smooth. Juvenile stromb shells in which the outer lip is not yet flared or thickened can easily be mistaken for cone shells because of their similar shape.

Strombs are herbivores; they feed on algae or detrital material. They are found on sand or mud flats or in sand patches on rocky or coral reefs, and they usually occur in great quantity in areas suitable for them. Often the shells are covered with algal growths and the collector needs a keen eye to spot them crawling on the sand, although the criss-crossed tracks they leave usually betray their presence.

The egg-masses are long jelly tubes covered with sand grains and coiled and twisted into a knotted tangle. The eggs are very tiny and numerous (one egg-mass may contain nearly half a million) and the embryos hatch as free-swimming veliger larvae.

Members of the genus *Lambis* are commonly known as spider and scorpion shells. *Lambis* has only 9 living species, all confined to the tropical Indo-West Pacific. There are 5 species in northern Australia. The soft parts of these animals are similar to those of the strombs but the shell has long spikes on the outer lip, the feature from which their common name is derived. Sexual dimorphism is an impressive feature of some members of this genus. In *Lambis (Lambis) lambis* the males are usually smaller than the females and the spikes are short and hooked towards the posterior end, whereas in females the spikes are long and curl upwards. Male shells of the Pacific subspecies of *Lambis (Harpago) chiragra* are smaller than those of the females and the aperture does not have the deep rose colour typical of the female shell. Scorpion shells live among coral rubble and algae on intertidal reefs and at depths of a few fathoms.

The genus *Terebellum* has only a single species although it is so variable that it has been given several different specific names. It is sometimes known as the Little Auger Shell. The shell is long and slippery-smooth and so is well adapted for burrowing in its sand habitat. It has a simple outer lip with no spines and does not resemble the shells of other members of the family, but the anatomy of the animal indicates that there is a close relationship.

The closely related *Tibia* and *Rimella* are the other genera in the family Strombidae. Each is characterized by a long, sometimes spike-like anterior canal and a deeply grooved and drawn out posterior canal. Both genera contain only a few species and are confined to the tropical Indo-West Pacific region. So far no species of *Tibia* has been reported from Australian waters but 1 species of *Rimella* is found on the north coast of Western Australia.

Selected references:
ABBOTT, R. TUCKER (1960). The genus *Strombus* in the Indo-Pacific. Indo-Pacific Mollusca, 1 (2): 33-146, pls 11-117.
ABBOTT, R. TUCKER (1961). The genus *Lambis* in the Indo-Pacific. Indo-Pacific Mollusca, 1 (3): 147-174, pls 118-134.
JUNG, PETER and ABBOTT, R. TUCKER (1967). The genus *Terebellum* (Gastropoda: Strombidae). Indo-Pacific Mollusca, 1 (7): 445-454, pls 318-327.

1. *Lambis (Lambis) lambis* LINNAEUS, 1758

Spire moderately high, pointed. Outer lip thickened, widely flared, bears 7 long slender spines, and a deep stromboid notch. Body whorl rough with spiral striae and heavy shoulder nodules. Shoulders of the spire whorls sharply angulate. Aperture and columella smooth except for a few weak lirae posteriorly, columella smooth; columella and ventral surface of body whorl glazed. Exterior colour variable, white-cream with brown or bluish patches, interior rich pink, orange or purple-tan.

20 cm. Common. Indo-West Pacific; Barrow Is., W.A. to central Qld.

2. *Lambis (Lambis) truncata* HUMPHREY, 1786

Spire moderately high (see remarks below). Outer lip widely flared and bears 7 long arched spines. Weak shoulder nodules and spiral ribs adorn the whorls. Aperture smooth except for a few lirae at the anterior end of the outer lip. Posterior canal forms a curved groove running across the underside of the spire to the base of the most posterior spine. Ventral surface heavily glazed. Exterior white-cream sparsely speckled with brown, deep interior white, becoming pink or purple-tan toward the edges of the aperture in mature specimens.

35 cm. This is the largest member of the genus. Moderately common. Indo-West Pacific; north-eastern Qld. The form of this species found in Australian waters is the subspecies *L. t. sebae* KIENER, 1843 (2) which has a pointed apex. The nominate subspecies is found in the western Indian Ocean (excluding the Red Sea), and in that form the spire apex is truncate, not pointed.

3. *Lambis (Lambis) crocata* LINK, 1807

Similar to *L. lambis* but smaller, with longer spines. Aperture solid orange and entirely smooth (i.e. no posterior lirae), shoulders of the spire whorls with a beaded keel, and the second, third and fourth spines are hooked towards the posterior.

15 cm. Moderately common. Indo-West Pacific; north-eastern Qld.

4. *Lambis (Millepes) scorpius* LINNAEUS, 1758

Outer lip thickened, widely flared and bears 7 flattened, knobbed spines. Body whorl rough with heavy nodules on the shoulders and 3 spiral rows of weaker nodules anterior to the shoulders. A beaded spiral keel present on the shoulders of the whorls on the short spire. Aperture strongly lirate on both sides, lirae extend well out on to the ventral surface of the body whorl. Exterior cream or fawn with light brown patches and bands, interior purple with white lirae, brown or orange near the margins.

17 cm. Moderately common. Central Indo-West Pacific; north-eastern Qld.

5. *Lambis (Harpago) chiragra* LINNAEUS, 1758

Outer lip flared and bears 4 broad, hollow, backward-curving spines on its margin in addition to long, pointed anterior and posterior canals which both point out to the left side. Body whorl rough with heavy nodules on the shoulders and spiral rows of smaller nodules more anteriorly. Aperture constricted and lirate. Posterior canal a deep, narrow, winding channel almost closed by folds of parietal and labial walls. Exterior white with brown patches and flecks, aperture rose-pink or orange, parietal wall glazed, orange-violet or cream.

25 cm. Common. Indo-West Pacific; North-West Cape, W.A. to north-eastern Qld. Male specimens are usually smaller and have stronger lirations on the columellar side of the aperture. The Australian form is of the nominate subspecies. Another subspecies, *L. c. arthritica* RÖDING, 1798, is found in the western Indian Ocean.

Plate 14. Strombidae ($\frac{9}{10}$ × natural size) ⬐

Plate 15. *Lambis (Lambis) lambis* LINNAEUS, 1758, Qld., showing long eye stalks. *(Photo*—Don Byrne).

1. *Strombus (Tricornis) sinuatus* HUMPHREY, 1786

Rather flat-based and high-spired. Outer lip rather thin, flat, thickened at the margin, widely flared, and bears 3 or 4 thin, flat, projecting blades at its posterior edge. Body whorl smooth except for a high central dorsal knob and 2 other smaller shoulder nodules. Heavy shoulder nodules also adorn the spire whorls. Aperture smooth. Exterior cream with orange-brown irregular patches which become spiral bands near the lip, interior rich purple-brown becoming pink or orange near the margin of the aperture.

13 cm. Common. Central Indo-West Pacific; reported from North-West Cape, W.A. and north-eastern Qld. The illustrated specimen is from the Philippines.

2. *Strombus (Labiostrombus) epidromis* LINNAEUS, 1758

Rather light for its size, with a moderately high spire. Outer lip slightly thickened and widely flaring. Body whorl smooth except for weak spiral lines anteriorly and a few low, elongate nodules on the shoulders. Spire whorls with small, distinct shoulder nodules. Aperture smooth, columella calloused. Exterior white or cream, with faint flecks of yellow-brown, interior glossy white.

9 cm. Said to be common at some localities. Central Indo-West Pacific; North-West Cape, W.A. to north-eastern Qld.

Plate 16. *Strombus (Aurisdianae) aratrum* RÖDING, 1798, Qld. *(Photo*—Don Byrne).

3. *Strombus (Lentigo) lentiginosus* LINNAEUS, 1758

Solid, with large protruding nodules on the shoulders, especially on the body whorl, and spiral rows of smaller nodules elsewhere on the body whorl. Outer lip only slightly flared, anterior canal short, posterior canal represented by a deep channel along the side of the spire. Aperture and columella smooth, columella projecting anteriorly. Exterior white with large chestnut spots in front of shoulder nodules, and mottled with green-grey or grey-brown, interior orange, becoming cream toward the margins, ventral surface and columella glazed.

10 cm. Common. Indo-West Pacific; Dampier Archipelago, W.A. to north-eastern Qld.

4. *Strombus (Lentigo) pipus* RÖDING, 1798

Solid and rather quadrate. Outer lip thickened but only slightly flared. Heavy shoulder nodules and several spiral rows of smaller nodules present on the body whorl. Spire low, spire whorls nodulose. Inner side of the outer lip strongly lirate, but columella smooth and calloused and the ventral surface of the body whorl heavily glazed. Exterior white or cream with orange-brown patches, interior dark purple, columella cream.

7 cm. Common in restricted areas. Indo-West Pacific; north-eastern Qld.

5, 5a. *Strombus (Conomurex) luhuanus* LINNAEUS, 1758

Solid, cone-shaped, spire low. Outer lip slightly inturned instead of flaring. Body whorl smooth, spire whorls with axial ribs. Anterior canal a deep siphonal notch. Periostracum thick, rough, green-brown. Exterior (when periostracum removed) white with bands of orange-brown, aperture smooth, rich orange or red, columella black-brown.

7 cm. Abundant. Central Indo-West Pacific; Qld. to northern N.S.W.

6, 6a. *Strombus (Euprotomus) aurisdianae* LINNAEUS, 1758

Solid and high-spired. Spire and body whorl heavily nodulose at the shoulders, with many fine spiral ribs on the body whorl, plus 1 or 2 stronger nodulose ribs. Outer lip thickened, flared, and ends posteriorly in a backward pointing projection. Anterior canal long and curved upwards at an angle of about 75 degrees. A few weak posterior lirae present in the aperture. Ventral side of the body whorl and the lower spire whorls glazed dark brown or orange, exterior mottled cream, brown and blue-grey, interior rich orange.

8 cm. Common. The nominate subspecies *S. aurisdianae aurisdianae* LINNAEUS, 1758 (6) is widely distributed in the Indo-West Pacific while the subspecies *S. a. aratrum* RÖDING, 1798 (6a) is confined to muddy mainland waters of north-eastern Qld. *S. a. aurisdianae* is also reported from north-eastern Qld. but from clear offshore coral-water. The relationship between the 2 subspecies is problematical.

7, 7a. *Strombus (Euprotomus) vomer* RÖDING, 1798

Very like *S. aurisdianae* but the anterior canal and the posterior projection of the lip are shorter, and the aperture and anterior end of the columella are lirate. Exterior similarly mottled cream and brown but the interior whitish, becoming orange, tan or sometimes almost black near the margins and on the columella.

8 cm. Common. The nominate form of this species is found only in the Ryukyu Islands and New Caledonia, but the subspecies *S. vomer iredalei* ABBOTT, 1960 (the illustrated form) is found in northern Australia from Shark Bay, W.A. to the N.T.

8. *Strombus (Doxander) vittatus* LINNAEUS, 1758

Moderately solid with a very high spire and flared outer lip. Body whorl smooth except for a number of spiral striae anteriorly, but spire whorls bear prominent axial ribs. Weak lirae present in the aperture. Columella slightly calloused and smooth. Exterior pale yellow-brown, sometimes with flecked darker spiral bands or faint zigzag lines, interior glossy white.

8.5 cm. Moderately common. Central Indo-West Pacific; north-eastern Qld.

9, 9a. *Strombus (Doxander) campbelli* GRIFFITH & PIDGEON, 1834

Solid, with a high spire and flared outer lip. Body whorl smooth except for spiral striae anteriorly and elongate nodules on the shoulders. Strong axial ribs present on the spire whorls, and deep spiral grooves behind the shoulders. Aperture sometimes weakly lirate but columella smooth and calloused. Exterior white with spiral bands of brown zigzag lines, flecks or blotches, interior white.

7 cm. Abundant. Northern Australia from Houtman Abrolhos, W.A. (sub-fossil at Fremantle) to N.S.W.

Plate 17. Strombidae (⁹⁄₁₀ × natural size)

Strombidae (Cont.)

1. *Strombus (Laevistrombus) canarium* LINNAEUS, 1758
Heavy, broad, with angular shoulders and a widely flared outer lip. Spire moderately high and tapers to a sharp point; with weak axial ribs and varices. Body whorl smooth except for a few striae anteriorly, aperture and columella smooth. Exterior white, grey or yellow-brown usually with a reticulate pattern of brown lines, interior white.
9 cm. Abundant. Central Indo-West Pacific; east coast of Qld.

2, 2a, 2b. *Strombus (Canarium) urceus* LINNAEUS, 1758
Solid with a high turreted spire and a thickened but only slightly flared outer lip. Shoulders heavily nodulose, fine spiral riblets present at the anterior end and near the lip. Axial folds sometimes present on the ventral side. There are numerous fine, irregular spiral lirae in the aperture. Columella calloused and lirate at the anterior and posterior ends. Exterior brown, white, cream or greenish with darker spiral lines, bands or blotches, deep interior white, purple or almost black, columella and outer lip margin white, interior and exterior of anterior canal tip black.
5 cm. Abundant. The nominate form is restricted to the central Indo-West Pacific region (e.g. (2) a specimen from Batangas Philippines). The Australian subspecies is *S. urceus orrae* which is common from North-West Cape, W.A. to the Gulf of Carpentaria (2a, 2b). It is more quadrate than *S. urceus urceus*, has a shorter anterior canal, and fewer and much larger shoulder nodules.

3. *Strombus (Canarium) labiatus* RÖDING, 1798
Solid, elongate, spire moderately high, shouldered and nodulose. Outer lip thickened, only slightly flared. Body whorl with prominent elongate nodules on the shoulders and weak spiral lines anteriorly and near the lip. Shoulder nodules become axial folds on the ventral side. Inner surface of outer lip with many fine spiral lirae. Columella calloused and crossed by fine lirae throughout its length. External colour variable, usually white, green or grey-blue, with brown or orange bands or patches, inner side of outer lip violet or orange, columella yellow or orange.
5 cm. Abundant. Central Indo-West Pacific; eastern Qld. There are also questionable records from Dampier Archipelago, W.A. *S. urceus* resembles this species but lacks lirae at the centre of the columella.

4. *Strombus (Canarium) mutabilis* SWAINSON, 1821
Solid, with a rather low spire, a thickened but only slightly flared outer lip, and broad nodulose shoulders. Body whorl smooth except for spiral striae anteriorly and sometimes a weak spiral cord around the centre. Many fine crowded spiral lirae present in the aperture, columella calloused and crossed by well marked lirae throughout its length. External colour extremely variable, usually bright cream with patches, spots or bands of yellow, brown, black, or orange, interior usually pink.
4 cm. Abundant. Indo-West Pacific; Cape Leeuwin, W.A. to central N.S.W. *Strombus wilsoni* ABBOTT, 1967 is very like this but there are strong axial riblets on the spire whorls, 5 to 7 elongate nodules on the shoulder of the body whorl and the columella is smooth at the centre. This species has been described from Zanzibar, Fiji, and from Onslow and Dampier Archipelago, W.A.

5. *Strombus (Canarium) erythrinus* DILLWYN, 1817
Elongate, with a high, shouldered and nodulose spire and a thickened but weakly flared outer lip. Strong spiral riblets encircle the whorls. Strong axial folds present on the spire whorls and the ventral side of the body whorl, and many fine lirae in the aperture and along the inner edge of the columellar callus. Exterior white or cream, mottled or banded with brown, aperture dark purple-brown, outer half of columella white.
5 cm. Common. Indo-West Pacific; north-eastern Qld.

Fig. 11. *Strombus (Dolomena) variabilis* SWAINSON, 1820. At a depth of 6 feet, Langford Island, Qld. (*Photo*—Neville Coleman).

6, 6a. *Strombus (Dolomena) plicatus* RÖDING, 1798
Small and solid and characterized by the shouldered spire and strongly thickened folded and widely flared outer lip. Strong axial folds present on the spire whorls and on the left and ventral sides of the body whorl diminishing to shoulder nodules on the upper part of the body whorl. Fine spiral riblets encircle the whorls. Columella thickly calloused, smooth at centre but lirate posteriorly and anteriorly, aperture with strong bifurcate lirae. Exterior cream or pale brown and faintly banded, interior purple-brown, columella white or orange.
5 cm. Uncommon. The nominate form is found in the Red Sea. Recently, specimens (6, 6a) of the subspecies *S. plicatus pulchellus* REEVE, 1851 were dredged between Onslow and the Dampier Archipelago, W.A. This subspecies was previously known from the Central Indo-West Pacific region.

7. *Strombus (Dolomena) dilatatus* SWAINSON, 1821
Rather thin, with a moderately high spire and a thin widely flared outer lip. Body whorl glossy but sculptured with minute spiral striae and weak shoulder nodules. Spire whorls axially plicate. Aperture strongly lirate but the calloused columella bears only a few obsolete lirae at the ends. Posterior canal groove deep, curves across the ventral surface of the spire. Exterior cream with pale brown blotches and bands, interior and columella white except for a purple-brown patch deep within the aperture.
6 cm. Uncommon. Central Indo-West Pacific; North-West Cape, W.A. to north-eastern Qld. The newly discovered Western Australian form (7) may be the subspecies *swainsoni* REEVE, 1850 for which the locality was previously unknown.

8, 8a. *Strombus (Dolomena) variabilis* SWAINSON, 1820
Rather thin, with a moderately high, shouldered, nodulose spire, and a widely flared outer lip notched posteriorly as well as anteriorly. Body whorl smooth except for spiral striae anteriorly and shoulder nodules which become elongate on the ventral side. Aperture and calloused columella smooth. Exterior white with 5 or 6 spiral bands of wavy brown lines or patches, interior glossy white.
6 cm. Common. Central Indo-West Pacific; N.T. and north-eastern Qld. The absence of internal lirae distinguishes this species from *S. vittatus* and *S. campbelli* which it resembles superficially.

9, 9a. *Strombus (Gibberulus) gibberulus* LINNAEUS, 1758
Moderately solid. Spire low, outer lip only slightly thickened and flared. Penultimate whorl characteristically swollen and distorted. Body whorl smooth and curiously flattened; spire whorls bear fine spiral striae and white varices. Lirae present in the aperture, columella calloused, smooth. Exterior white with spiral bands of brown blotches, interior white, or tinted with violet, brown or yellow. The columella often has an elongate brown patch.
6 cm. Abundant. Western Pacific including Qld. and the coast of W.A. as far south as Barrow Is. The Australian subspecies is *S. gibberulus gibbosus* RÖDING, 1798 (9, 9a).

10, 10a. *Rimella cancellata* LAMARCK, 1816
Solid and fusiform. Spire high and attenuate, outer lip thick but weakly flared, anterior canal projects forward almost horizontally. Sculpture strong, consists of axial folds with fine spiral riblets in the interspaces, and usually 2 varices on each of the spire whorls. Long posterior canal groove and runs backwards over 2 or more of the spire whorls on the right side. Inner edge of outer lip lirate, columella calloused and smooth. Exterior mauve or yellow-brown between the white or pale yellow axial folds, with narrow yellow-brown bands, aperture mauve, columella white.
3.5 cm. Uncommon. Central Indo-West Pacific; Dampier Archipelago (the illustrated specimens) to Broome, W.A.

11, 11a. *Terebellum terebellum* LINNAEUS, 1758
Rather thin, slender and cylindrical, with deeply channelled sutures, and without a stromboid notch. Aperture long, narrow, and broadly open at the anterior end. Outer lip slightly thickened but not flared. Exterior glossy smooth, colour and pattern extremely variable, usually with a pale yellow, brown or cream background and red-brown bands, zigzag lines, spiral lines, dots or flecks.
7 cm. Common. Indo-West Pacific; North-West Cape, W.A. to north-eastern Qld. The many colour forms have been named separately but a recent study has shown that they are all one species.

Plate 18. Strombidae ($1\frac{7}{10}$× natural size)

Plate 19. *Cypraea saulae* GASKOIN, 1843, Qld.   *(Photo*—Don Byrne).

cowries (FAMILY CYPRAEIDAE)

PROBABLY NO OTHER GROUP of molluscs attracts such universal attention as the cowries. Enthusiasts pay high prices for some of the uncommon species, but few collections lack representatives of the family. The attraction of the shells lies in their high gloss, striking colours and colour patterns, and graceful shapes.

Estimates of the number of living species vary according to the working philosophy of the taxonomist. One recent list claims 185 living species but more conservative authors recognize about 160. Most of these species have been divided into 2 or more subspecies. The family also has been divided into many genera and subgenera but this is at present a controversial matter and in this book only a single generic name, *Cypraea*, is used with few subgeneric divisions.

A large proportion of living cowries is found in the tropical seas of the Indo-West Pacific region. Species are widespread there, many of them occurring in northern Australia. Only a few species inhabit the Mediterranean and Caribbean Seas, and the tropical west coast of Central America. The temperate waters of southern Australia support a number belonging to especially interesting and distinctive groups

(ranked as subgenera here). All of these groups have a long fossil record in sediments of southern Australia and they seem to have been a distinctive feature of the fauna of the southern Australian Region for many millions of years.

1. The subgenus *Zoila.*
These very attractive medium to large sized cowries living along the coasts of South and Western Australia are among the most sought after of all Australian shells. Opinions differ on the status of the group and the taxonomic treatment of the species. In this book we follow the conservative approach adopted by Wilson and McComb (1967) who rank *Zoila* as a subgeneric group and recognize only 5 species as follows: *friendii, venusta, marginata, rosselli* and *decipiens*. The centre of distribution of the subgenus seems to be the south-western corner of Western Australia where 4 of the 5 species are found. Of these 2 also extend into South Australian waters. One of them, *(decipiens)* is found on the tropical shores of northern Western Australia. Fossil species are known from Tertiary rocks of south-eastern Australia, and one of these, *C. (Zoila) gigas* McCoy, 1867 reached a length of almost 1 foot.

2. The subgenus *Notocypraea.*
So far there has been no general agreement concerning the taxonomy of the small, thin-shelled, extremely variable cowries which constitute this group. 6 species are described in this book: *molleri, declivis, pulicaria, piperita, angustata* and *comptoni.* Other names have been used which may represent valid species, e.g. *dissecta* IREDALE, 1931; *emblema* IREDALE, 1931; *euclia* STEADMAN & COTTON, 1946; *wilkinsi* GRIFFITHS, 1959. For 2 recent but contrasting taxonomic arrangements of *Notocypraea* see Griffiths (1962) and Schilder (1964). Most of the species are relatively common under stones in the intertidal zone and down to depths of several fathoms.

3. The subgenus *Umbilia.*
This group of large cowries flourished in the southern Australian Region during the Tertiary period and there are many eye-catching fossil species. However, only 2 species *(hesitata* and *armeniaca)* have survived to the present day. Both are deep water animals taken only by trawlers or in craypots. Nothing is known of their natural history. Shells of this group are remarkable for their sunken spires and peculiarly flattened and extended anterior ends.

4. The subgenus *Austrocypraea.*
A solitary living species *(reevei)* represents this group although there are similar fossil shells in mid-Tertiary sediments. The affinities of this subgenus with other cowries is still in doubt.

Cowry shells have an unusual form and people often overlook the fact that they are spirally coiled like other gastropod shells. The difference is that the coils overlap so that each covers the one beneath it (Fig. 12). The aperture extends almost the full length of the shell and, in the adult stage, it is bordered by raised transverse ridges or teeth which are a characteristic feature of cowries (although similar teeth are found in a few other shells).

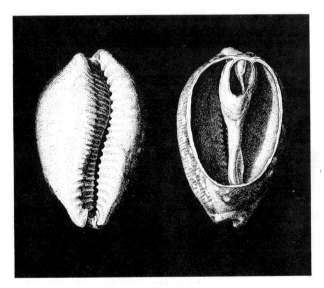

Fig. 12. *Cypraea vitellus* LINNAEUS, 1758.
Left: Under surface showing teeth.
Right: Sectional view showing overlapping coils.

The juvenile cowry shell is often mistaken for an olive or some other kind of shell because it is thin, rather long and slender, and has a pointed spire and a sharp untoothed outer lip (Fig. 13). It grows in the usual manner of gastropod shells by additions to the outer lip, as a bricklayer builds a wall, but spirally instead of straight. The shell thus becomes wider, longer and more voluminous. But when the animal nears sexual maturity and the gonads develop, the last body whorl becomes swollen and spacious and covers over the spire. At the same time the outer lip turns inwards so that the aperture is narrowed to a mere longitudinal slit. The shell does not grow in length after that but is thickened by deposition of shell material over the whole surface, and at this late stage the apertural teeth are formed.

Fig. 13
Juvenile *Cypraea arabica*
LINNAEUS, 1758

Some authors have reported that the adult animal is capable of dissolving its shell and growing a new one but this has not been satisfactorily documented, and seems very unlikely. However repairs are often made to broken shells by the same methods used in shell growth and thickening.

The high gloss polish of cowry shells (the French name for them is "porcelaine") is produced by the animal's bilobed mantle which spreads up and over the shell from both sides. Living cowries are usually found with the mantle lobes extended to cover the shell, but these delicate structures are quickly withdrawn into the aperture when danger threatens. In many species the mantle bears branching filaments or papillae of uncertain function which provide useful characters for taxonomic work.

Cowries feed most actively at night. The feeding habits are known for only a few species but it is probable that most browse generally on tiny animals or plants that grow on rock and coral. Species of the subgenus *Zoila* of Western and South Australia feed on sponges which are sedentary animals. So these cowries can be regarded as both browsers and carnivores. There have been reports that in aquaria some cowries will eat the bodies of worms or crushed bivalves (pers. comm. John Orr) which suggests that they may be predators in natural conditions, although this idea is not supported by the simplicity of the radula and mouth parts. This matter needs additional careful field and laboratory study.

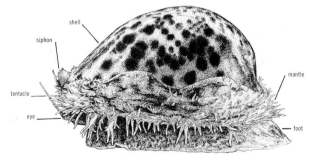

Fig. 14. *Cypraea tigris* LINNAEUS, 1758, showing the major external features of the body.

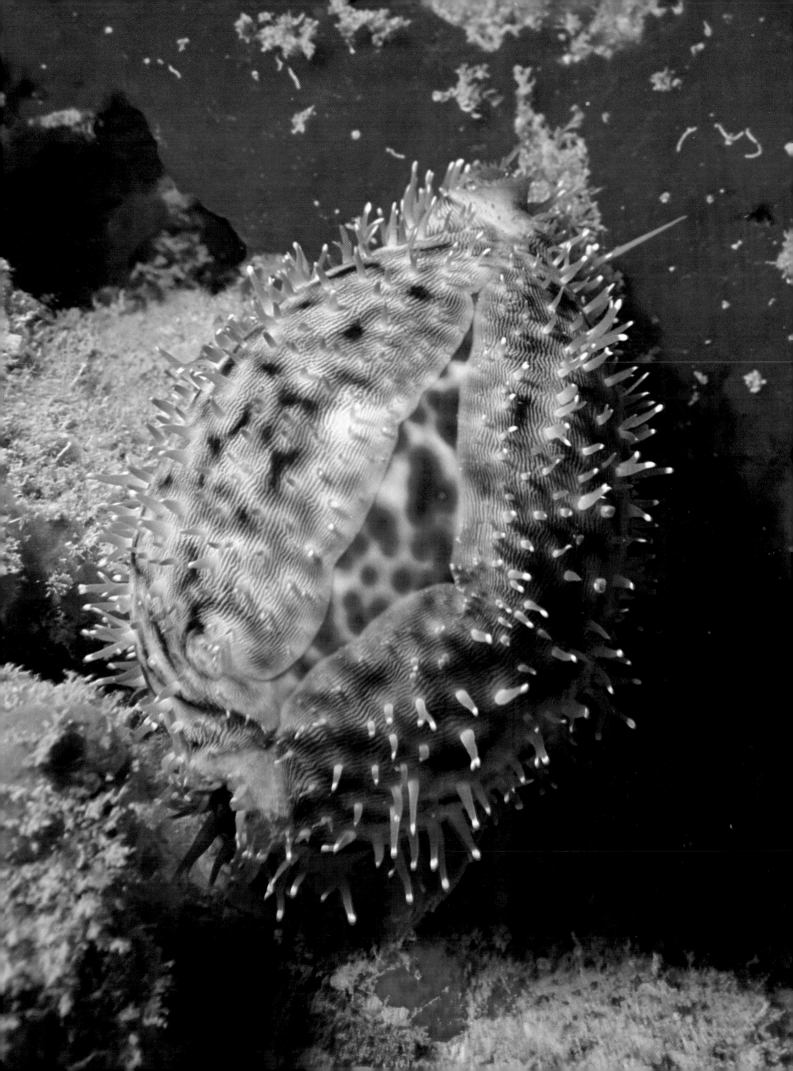

Plate 21. *Cypraea staphylaea* Linnaeus, 1758, Qld.
(*Photo*—Don Byrne).

Plate 22. *Cypraea punctata* Linnaeus, 1771, with eggs, Qld.
(*Photo*—Barry Wilson).

The radula of cowries is simple and consists of several hundred rows of teeth, with 7 teeth in each row. Details of the shape of the radular teeth may be useful taxonomic characters, but in some species they vary as much as the shells do.

In cowries the sexes are separate and the animals copulate to achieve fertilization of the eggs within the female reproductive tract. The female lays her circular or oblong egg-masses under stones or in crevices. Each consists of several hundred capsules glued together in a gelatinous mass. Each capsule contains 100-300 eggs, but in some species only 1 of these develops to become an embryo. In most species the young hatch from the eggs as veliger larvae which lead a planktonic life for a brief period (usually a few days) before settling to the bottom. A few species (e.g. some species of the subgenus *Notocypraea*) hatch directly from the eggs as crawling juveniles and there is no pelagic larval stage. Most cowries may be found alive under stones or corals in the intertidal zone. Pure coral reefs are not usually good collecting sites; a rocky reef with some corals growing on it will often provide better results. As many as 30 different species may be collected on some reefs in northern Australia. Many deep water forms are known and some African species have only been found in the stomachs of large fish. Species of the subgenus *Zoila* do not hide under stones but live out in the open on sponges or on the walls of underwater caves.

Selected references:

Allan, Joyce (1956). *Cowry Shells of World Seas.* Georgian House, Melbourne, X + 170 pp.

Burgess, C. M. (1970). *The Living Cowries.* New York.

Cate, Crawford, N. (1964). Western Australian Cowries (Mollusca : Gastropoda). Veliger, 7 : 7-29, 5 pls.

Cate, Crawford, N. (1968). Western Australian Cowries—a second, revised and expanded report. Veliger, 10 : 212-232; pl. 21—34, 5 maps.

Griffiths, R. J. (1962). A review of the Cypraeidae genus *Notocypraea*. Mem. natn. Mus. Vict., no. 25 : 211-231, pls 1-3.

Schilder, F. A. (1964). Provisional classification of the genus *Notocypraea* Schilder, 1927 (Cypraeidae). Veliger, 7 : 37-43.

Schilder, F. A. & Schilder, M. (1938-39). Prodrome of a monograph on living Cypraeidae. Proc. malac. Soc. London, 23 (3-4): 119-231.

Wilson, B. R. & McComb, Jennifer A. (1967). The genus *Cypraea* (Subgenus *Zoila* Jousseaume). Indo-Pacific Mollusca, 1 (8): 457-484, pls 329-344.

Wilson, B. R. & Summers, Ray (1966). Variation in the *Zoila friendi* (Gray) species complex (Gastropoda : Cypraeidae) in south-western Australia. J. malac. Soc. Aust. 1 (9): 3-24, pls 1-4.

(For shells described below see Plate 25, page 47).

1, 1a, 1b, 1c, 1d, 1e, 1f, 1g. *Cypraea (Zoila) friendii friendii* Gray, 1831

(see also colour plate of live animal, page 13, plate 5)

This species is so variable that it is difficult to give a concise description that fits all specimens. Any single population contains much individual variation but there are also clinal trends, geographic variants and localized environmental forms (ecophenotypes). Most significantly, the width of the shell, relative to its length, increases from north to south and east in a clinal manner. In addition, there is a strong tendency for shells from deep water to be wider than shells from shallow water in the same region. It has been suggested that this two-way clinal increase in shell width may be related to decreasing water temperature.

Several of the variants of this species have been given names but a detailed study of large series of specimens over the entire range of the species has shown that these names are of doubtful value as formal taxa (Wilson and Summers, 1966). In our opinion these names should not be used as subspecies names because doing so implies much about the genetic relationships of the variants which does not appear to be in accordance with available evidence. However, collectors may find it useful to use them as *forma* when referring to specimens characteristic of a local population or habitat type. The forms may be listed and described as follows:

(i) Typical *C. friendii friendii* (1b, 1d, 1e).

The type locality is near Fremantle, W.A. (1b) and specimens from shallow water in that area are elongate, with a projecting spire (especially in sub-adults) and sharp edged up-turned sides to the anterior and posterior canals. The labial teeth are short and fairly strong but there are only a few weak anterior teeth on the columella. The colour of the sides and base is chocolate brown; the top is cream with 4 bands of pale blue blocks overlain to varying degrees by irregular brown blotches. The sides have large brown spots. Aberrant golden shells are known from several localities.

Further south in Geographe Bay, W.A. the shells are similar but the average width is greater (1e). A dwarf form occurs in shallow water at Dunsborough and Quindalup, Geographe Bay, W.A. (1d).

(ii) *C. friendii* forma *vercoi* (1c).

Shells from the southern coasts of W.A. between Eucla and Cape Leeuwin are much broader than those from shallow water on the central west coast, and they have a wider aperture. Specimens from Esperance were given the name *vercoi* (Schilder 1930) but the status of this has been questioned because of the clinal variation of shell width.

(iii) *C. friendii* forma *jeaniana* (1, 1a).

Broad shells also occur in 20-60 fathoms off the west coast of W.A. between Fremantle and Carnarvon, but these have other characters which distinguish them from the typical shallow water forms of the Fremantle region. They are rather flat-based and have weak teeth along the full length of the aperture on the columellar side. The extreme northern end-of-range population off Carnarvon has been given the subspecies name *jeaniana* (Cate, 1968) but specimens from slightly more southern localities show less divergence from the typical form and the name seems to be useful only as a form name. (1) is from 60 fathoms in Shark Bay; (1a) is from 20 fathoms off False Entrance.

(iv) *C. friendii* forma *contraria* (1f, 1g).

From along the outer edge of the continental shelf off the southern coast of W.A. at depths of 60 to 110 fathoms extremely pale shells have been trawled. These specimens are even more broad and tumid than the shallow water *vercoi* form of the south coast. Pure white specimens are known but more often the base is pale orange and there are large pale orange or tan spots along the sides and smaller spots and blotches on the dorsum (1g). Iredale (1935) gave these the name *contraria* supposing them to represent a deep water, pale western subspecies of the South Australian "species" *C. thersites*.

Since then similarly coloured shells have been trawled at similar depths off the west coast of W.A. between Fremantle and Geraldton (1f). These west coast pale shells have the shape and dentition of the *jeaniana* form. So it seems that there may be pale-shelled populations belonging to the *C. friendii* complex along the outer edge of the continental shelf throughout most of the species range. The lack of colour is almost certainly due to environmental effect associated with depth and these colour forms probably deserve no taxonomic standing. Much more information is needed about deep water forms, and until that is available we hope that students will refrain from adding more names and more confusion to this complex situation.

10 cm. Moderately uncommon. The full range of the species is from Eucla to Carnarvon, W.A. at depths from 1 to 110 fathoms. A moderately large population which once lived at the type locality in Cockburn Sound, near Fremantle has been heavily "over-fished" by unscrupulous collectors in recent years.

2, 2a. *Cypraea (Zoila) friendii thersites* GASKOIN, 1849

Broad, flat-based, base pale-coloured. Many specimens have faint teeth at the extreme posterior end of the columella. Spire short, sides of anterior and posterior canals low, not elevated (characters related to the tumidity of the shell).

9 cm. S.A. This eastern population appears quite isolated from the W.A. *C. friendii friendii* population. Shape and base colour are consistent characters which distinguish the shells from the much more variable western shells, and taxonomic distinction for the S.A. population is definitely necessary. However, because of the clinal increase in shell width from north to south and east, possibly associated with decreasing water temperature, and the fact that there are no apparent anatomical differences (Wilson & Summers, 1960), we believe that the S.A. population warrants separation only at subspecific rank. Other authors give *thersites* full specific rank.

3, 3a, 3b, 3c. *Cypraea (Zoila) venusta* SOWERBY, 1846

Another variable species. Broad, subpyriform, with rounded margins and depressed spire. Weak teeth at the anterior and posterior ends of the columella, and sometimes along its whole length. Two colour forms known. The original description was based on a very pale, creamy-pink or creamy-yellow shell with pale brown spots. The more common form has a pale base but a dark brown top and grey sides with a fine granular pattern (like a "grainy" photographic print).

Within each of these two colour forms size and shape varies. Pale specimens from the west coast of W.A. between Fremantle and Shark Bay have the same shape as dark shells from the same area. Only a few pale specimens are known, some strongly spotted, others without spots, (3b) is an unspotted shell taken from a cray-pot off Kalbarri north of Geraldton. A pale form of rather different shape is known from 2 specimens trawled on the outer edge of the continental shelf in the western part of the Great Australian Bight (3). These shells are very tumid and rose-pink but their aperture dentition removes any doubt of their close affinity to the west coast form.

Each of 2 size forms of the dark west coast shells has a distinct name. In the Cape Leeuwin, Cape Naturaliste and Geographe Bay region the shells are large (over 7 cm in length). Iredale (1939) named this form *episema* (3c). Uniformly golden coloured specimens of this form from Geographe Bay seem to be aberrant shells and not true members of the pale form.

In shallow water at Sorrento Beach, Carnac, Garden and Penguin Islands, all close to Fremantle, is a population of smaller shells (average length 5.8 cm, maximum 6.5 cm). The shells of these populations are variable in colour. Some are very dark chocolate brown, others cinnamon. Schilder (1963) named this form *sorrentensis* (3a). However, it has been shown that, apart from size, the shell characters of the large and small forms are the same and that there are no differences in the anatomy or radula. Large dark shells also occur in deeper water off Rottnest Is. and at the Houtman Abrolhos ("Abrolhos Islands"). There seems little value in maintaining the large and small forms as distinct taxa. Schilder (1963) described a large aberrant specimen of distorted shape from the Abrolhos as a distinct species which he named *catei*, but this too seems dubious.

The pale and dark forms are very closely related, but whether they are distinct species, or forms of a single species, still remains to be determined. Because of their similarity in shape and aperture dentition we propose to regard them as forms of the 1 species for the time being.

8 cm. Dark specimens are moderately uncommon, but only a few specimens of the pale form are known. The total range of the species is from the western part of the Great Australian Bight to Shark Bay, W.A. at depths from 1 to 100 fathoms.

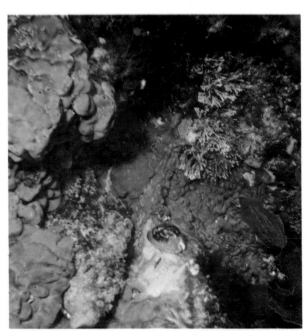

Plate 23. *Cypraea (Zoila) venusta* SOWERBY, 1846, feeding on a sponge in an underwater cave off Sorrento, W.A.
(Photo—Barry Wilson).

Plate 24. *Cypraea (Zoila) venusta* SOWERBY, 1846, on sponge. The small white patches next to the animal are holes it has bitten in the sponge, Sorrento, W.A. *(Photo*—Barry Wilson).

Plate 25. Cypraeidae *(Zoila)* (⁹⁄₁₀× natural size)

Cypraeidae (Cont.)

1. *Cypraea (Zoila) decipiens* SMITH, 1880

Subpyriform with an unusually high dorsal hump. Base flattened, margins rounded, teeth present on both sides of the aperture. Base and sides uniform chocolate brown without spots, top cream with varying amounts of irregular brown blotching and a curious faceted texture.

7 cm. Moderately common. Northern W.A. between North-West Cape and Buccaneer Archipelago. Taken by divers to 40 fathoms.

2, 2a. *Cypraea (Zoila) marginata* GASKOIN, 1848

Ovate with a fairly high dorsal hump, projecting spire, flat base and sharply angular irregularly indented sides. Teeth strong on both sides of the aperture. Base white with brown spots in depressions, sides heavily spotted, the top wholly white, or white with varying amounts of brown spots or blotches.

6 cm. S.A. and W.A. as far north as Carnarvon. The species was known from only 1 specimen until a few years ago but recently it has been collected in small numbers by skin-divers. Shells from S.A. and the south coast of W.A. are broader and more tumid than west coast shells.

3. *Cypraea (Zoila) rosselli* COTTON, 1948

Broad, with high dorsal hump and sharply angulated margins. Fairly strong teeth present on both sides of the aperture. Base, top and sides dark chocolate brown without spots, top usually with a large cream blotch.

6 cm. This species was described relatively recently from beach-worn specimens collected near Fremantle, W.A. The first live-taken specimen was taken in 1962 off Rottnest Is. by Mr Max Shaw when scuba-diving at 35 fathoms. Since then a number of living specimens have been taken, mostly in craypots, but the species is still one of the most difficult to obtain of all Australian shells. So far the species has been recorded from localities between Bunbury and Shark Bay, W.A. at depths between 3 and 40 fathoms.

4, 4a. *Cypraea argus* LINNAEUS, 1758

Solid and cylindrical, with spire whorls visible. Aperture almost straight, with strong teeth on the outer lip, and finer, more numerous teeth on the columella extending deep into the aperture. Base fawn with 2 large red-brown blotches. Fawn-grey top crossed by 2 pale grey bands, and prominently patterned with large irregular brown rings.

10 cm. Uncommon. Indo-West Pacific; North-West Cape, W.A. to central Qld.

5, 5a. *Cypraea talpa* LINNAEUS, 1758

Solid and subcylindrical, with a narrow, almost straight aperture, bordered by fine and numerous teeth. Base and sides very dark chocolate brown and lacking spots. Top light brown with 3 light creamy-brown bands.

9 cm. Uncommon. Indo-West Pacific; North-West Cape, W.A. to northern N.S.W.

6. *Cypraea mappa* LINNAEUS, 1758

Solid, pyriform to ovate, high and inflated. Margins pinched in at the anterior extremity. Aperture narrow, outer lip teeth strong and short, columellar teeth finer, more numerous and extend deep into the aperture. Base pinkish or purple-white; sides spotted; top pale brown with some clear patches between crowded brown irregular lines which tend to be longitudinal, and a conspicuous mantle line with short lateral extensions.

8 cm. Uncommon. Indo-West Pacific; Qld.

Plate 26. *Cypraea talpa* LINNAEUS, 1758, with eggs, Qld.
(*Photo*—Barry Wilson).

7. *Cypraea tigris* LINNAEUS, 1758

(see also colour plate of live animal, page 44, plate 20)

Very solid, ovate to slightly pyriform, with rounded margins. Outer lip teeth strong but short, anterior columellar teeth strong and extended into the aperture, but the posterior ones weak and short. Colour variable, shell usually white or pale brown with large dark brown spots on the top and sides. Mantle line present.

13 cm (larger in Hawaii). Common. Indo-West Pacific; Houtman Abrolhos, W.A. to northern N.S.W. Western and eastern Australian forms are recognized as distinct subspecies, i.e. *C. t. pardalis* SHAW, 1795 and *C. t. lyncichroa* MELVILL, 1888 respectively. The illustrated shell is from Barrow Is., W.A.

8. *Cypraea vitellus* LINNAEUS, 1758

Solid, swollen and ovate to pyriform. Strong teeth present on outer lip, columellar teeth fine, more numerous and extend deep into aperture. Base white or pale lilac, becoming brown toward the margins, sides and top brown with conspicuous white spots of varying size. Fine vertical clouded white striae characteristically present on the sides.

7 cm. Common. Indo-West Pacific; Cowaramup Bay, southern W.A. to central N.S.W. Two Australian subspecies are recognized, *C. v. vitellus* in the west (8) and *C. v. orcina* IREDALE, 1931 in the east.

9. *Cypraea mauritiana* LINNAEUS, 1758

Very solid, broad and flat-based, with angular margins and strong teeth. Base and sides dark chocolate brown, top slightly paler with large, more or less rounded cream spots.

11 cm. Common. Indo-West Pacific. Kimberley, W.A. to northern Qld. The northern form from W.A. is known as *C. m. regina* GMELIN, 1791, and the eastern form is *C. m. calxequina* MELVILL & STANDEN, 1899 (9).

10. *Cypraea testudinaria* LINNAEUS, 1758

Solid, elongate-cylindrical, with a straight, narrow aperture which widens just behind the anterior end. Outer lip teeth strong but short, columellar teeth weaker and extend deep into the aperture except posteriorly. Base light brown, sides spotted, top fawn with large brown spots, overlain by dense dark brown blotches and scattered minute pale grey dots.

13 cm. Uncommon. Indo-West Pacific. There are few reliable northern Australian records but there is 1 beach specimen in the W.A. Museum collection from the North Kimberley coast of W.A.

Plate 27. *Cypraea mappa* LINNAEUS, 1758. At a depth of 25 feet, Tryon Island, Qld. (*Photo*—Neville Coleman).

Plate 28. Cypraeidae ($\%_{10}$ × natural size)

Cypraeidae (Cont.)

1, 1a. *Cypraea eglantina* Duclos, 1833

Solid, elongate and subcylindrical, with a projecting spire, a rather flat base and slightly calloused but rounded margins. Teeth and colouring like those of *C. arabica* but there is a brown blotch beside the spire. Mantle line distinct.

8 cm. Common. Indo-West Pacific; Houtman Abrolhos, W.A. to eastern Qld. The W.A. form (1, 1a) has been given subspecific status as *C. e. perconfusa* Iredale, 1935. This species is commonly confused with *C. arabica* but can be distinguished by the presence of the spire blotch and more rounded margins.

2. *Cypraea (Umbilia) armeniaca* Verco, 1912

Very like *C. hesitata* but more globular, and less rostrate anteriorly. Base orange, sometimes with a golden-orange patch on the top.

10 cm. 80-100 fathoms off Eucla, on the south coast of W.A. Few specimens are known. This apparently isolated population may be only a western subspecies or form of *C. hesitata*.

3, 3a, 3b. *Cypraea (Umbilia) hesitata* Iredale, 1916

Large, pyriform, broad and rounded near the centre but with the anterior end rostrate. Spire deeply impressed (umbilicate), teeth along both sides of the aperture rather weak. Creamy-white with light brown spots on the sides, light brown smears and spots on the top, and darker brown smudges at the anterior and posterior ends.

10 cm. This species was once considered rare but is now trawled rather commonly off south-eastern Australia from Tas. and eastern Vic. to southern N.S.W. A deep water albinistic form from eastern Vic. (3a, 3b) has been named *howelli* Iredale, 1931 but this name is not usually recognized. Some N.S.W. shells are much smaller than those from the south and have been given the subspecies name *beddomei* Schilder, 1930 but that name is also doubtful.

4, 4a. *Cypraea (Austrocypraea) reevei* Sowerby, 1832

Rather thin and ovate or globose, with a rounded base and a narrow aperture bordered by numerous fine teeth which, on the columellar side, extend deep into the aperture. Surface of the shell covered with small dents or depressions as if hammered. Base white, sometimes tinted pink along sides of aperture, sides and top light brown, unspotted, but sometimes with slightly darker bands, ends pink.

4.5 cm. Rarely collected alive. The species is found in S.A. and southern W.A. as far north as the Houtman Abrolhos. Shells from the northern end of the range tend to be more globose than southern specimens.

Plate 29. *Cypraea eglantina* Duclos, 1833. At a depth of 10 feet, Tryon Island, Qld.   (*Photo*—Neville Coleman).

5. *Cypraea arabica* Linnaeus, 1758

Solid, ovate to subcylindrical, with a slightly projecting spire and a rather flat, broad base. Margins calloused, especially at the anterior end. Short, fairly strong teeth present on outer lip, but columellar teeth fine, more numerous and at the anterior end, extend deep into the aperture. Base creamy-grey, sometimes with a reddish blotch on the left side, teeth brown, top fawn-grey with a hieroglyphic pattern of brown markings, many tending to be longitudinal lines, interrupted by clear spaces. Large brown spots present on the sides, dark brown blotches at the ends, but there is no spire blotch (see *C. eglantina*). Mantle line prominent.

8 cm. Common. Indo-West Pacific; North-West Cape, W.A. to central N.S.W. Cate's (1964) separation of western Australian populations as a distinct subspecies *C. a. brunnescens* does not seem justified.

6. *Cypraea scurra* Gmelin, 1791

Solid, elongate and cylindrical, with a straight, narrow aperture, fine, numerous and short teeth, and rounded base and margins. Spire projects slightly but covered by callus. Base and sides fawn, teeth red-brown, sides with large brown spots. Top with a reticulate pattern of light brown lines which enclose large round blue-grey spaces. Mantle line distinct, spire blotch lacking.

5 cm. Uncommon. Indo-West Pacific; Qld. and northern N.S.W.

7. *Cypraea isabella* Linnaeus, 1758

Solid, nearly cylindrical, with a straight, narrow aperture bordered by very fine numerous teeth. Base and sides white and lack spots, top fawn or pale grey with irregular black lines which tend to be longitudinal, and orange blotches at the ends.

4 cm. Fairly common. Indo-West Pacific; Houtman Abrolhos, W.A. to central N.S.W. The eastern Australian subspecies is *C. i. lekalekana* Ladd, 1934. The western Australian subspecies is *C. i. rumphii* Schilder & Schilder, 1938.

8, 8a. *Cypraea lynx* Linnaeus, 1758

Solid and ovate or pyriform. Outer lip teeth strong and short, columellar teeth also strong but confined to the aperture. Base and sides white or fawn, bright orange between the teeth. Top pale blue-grey or fawn with numerous irregular clouded mauve spots and some large brown spots. Large brown spots present on the sides, dark brown blotches at the ends, and a prominent mantle line on the dorsum.

5 cm. Common. Indo-West Pacific; Houtman Abrolhos, W.A. to northern N.S.W. Two Australian subspecies are recognized, *C. l. vanelli* Linnaeus, 1758 in the west (8, 8a) and *C. l. caledonica* Crosse, 1869 in the east.

9. *Cypraea carneola* Linnaeus, 1758

Solid and ovate or slightly pyriform. Outer lip teeth strong and short, columellar teeth more numerous and extend well into aperture. Base and sides fawn, teeth deep lilac or purple, top light orange-brown with 4 darker bands.

5 cm. Common. Indo-West Pacific; Houtman Abrolhos, W.A. to northern N.S.W. W.A. shells are said to be typical of the species but Iredale (1939) gave the subspecies name *C. c. thepalea* to the eastern Australian form (9). *C. leviathan* Schilder & Schilder, 1937 (not illustrated) is like a large elongate form of *C. carneola* but is said to differ anatomically; the shell-collector will have difficulty separating these 2 species. Cate (1968) recognized the W.A. form of *C. leviathan* from North-West Cape as a distinct subspecies which he named *gedlingae*.

10, 10a. *Cypraea histrio* Gmelin, 1791

Solid and ovate with small, short teeth. Broad, margins calloused and expanded behind the anterior end. Base white or pale fawn, teeth brown, sides with large dark brown spots, top fawn or blue-grey with indistinct darker bands and a reticulate pattern of light brown lines. There are dark brown blotches at the ends.

8 cm. Common. Indian Ocean; North-West Cape, W.A. to Darwin. The Australian form has been recognized as a subspecies *C. h. westralis* Iredale, 1935.

Plate 30. Cypraeidae (⁹⁄₁₀ × natural size)

Cypraeidae (Cont.)

1. *Cypraea caurica* LINNAEUS, 1758

Solid, cylindrical, with unusually strong teeth. Base and sides pale brown, often slightly tinted purple. Large, widely spaced spots present on the sides, top greenish, densely freckled with minute and diffuse brown spots and crossed by 3 brown bands. Usually a large dark brown blotch present at the centre of the top, and conspicuous brown blotches on each side of the anterior and posterior channels.

4.5 cm. Common. Indo-West Pacific; Houtman Abrolhos, W.A. to N.S.W. Iredale recognized 2 Australian subspecies, *C. c. longior* IREDALE, 1935 in the east and *C. c. blaesa* IREDALE, 1939 in the west (1).

2, 2a. *Cypraea teres* GMELIN, 1791

Solid, subcylindrical, with sunken spire and calloused right margin. Outer lip teeth fairly strong, columellar teeth fine, numerous and confined to the aperture. Base white, sides white with large brown spots, 3 double bands of squarish brown spots cross the bluish-grey top, often with irregular V-shaped marks between the bands.

5 cm. Uncommon. Indo-West Pacific; Houtman Abrolhos, W.A. to central N.S.W. The western Australian form is said to be typical of the species but the eastern Australian form belongs to the subspecies *C. t. pentella* IREDALE, 1939.

3. *Cypraea erosa* LINNAEUS, 1758

Solid, flat-based, sides of the anterior end slightly pinched. Teeth very strong, those of the right side extend across the base almost as far as the margin. Base and sides yellow-white with a few interrupted brown lines near the margins, and a large brown blotch on each side, top olive-green with numerous small green spots and a few larger brown spots.

5 cm. Common. Indo-West Pacific; Houtman Abrolhos, W.A. to northern N.S.W. The Australian subspecies is *C. e. purissima* VREDENBURG, 1919.

4, 4a. *Cypraea ovum* GMELIN, 1791

Solid, pyriform. Very like and often confused with *C. errones* but has stronger columellar teeth, no trace of anterior terminal blotches, and the aperture less dilated anteriorly. Orange colouring between the aperture teeth most characteristic.

3 cm. Central Indo-West Pacific and Micronesia; Houtman Abrolhos, W.A. to central Qld. This may be only a form of *C. errones*.

5. *Cypraea errones* LINNAEUS, 1758

Solid, subpyriform. Outer lip teeth fairly strong and well spaced, columellar teeth rather weak. Aperture narrow posteriorly, dilated anteriorly. Base and sides creamy-white or yellow and unspotted, top green with 3 faint blue bands overlain by fine diffuse brown freckles, and usually a large central brown blotch. Large anterior terminal blotches or spots often present.

3.5 cm. Very common. Indo-West Pacific; Shark Bay, W.A. to northern N.S.W. The western Australian subspecies is *C. e. coxi* BRAZIER, 1872 and the eastern Australian subspecies is *C. e. coerulescens* SCHRÖTER, 1804 (5). Schilder (1968) described a blue variant of this species from Broome, W.A. as a new subspecies which he named *azurea*. However, this colour form is common among populations of the normally coloured shells and there are many intergrades.

Plate 31. *Cypraea cribraria* LINNAEUS, 1758, Qld.
(*Photo*—Don Byrne).

6. *Cypraea labrolineata* GASKOIN, 1848

Solid, elongate-ovate or subcylindrical. Margins thickened and pitted. Teeth strong, those on the outer lip extending almost to margin. Base white, sides white with conspicuous dark brown spots above the marginal rim, conspicuous dark brown blotches present on each side of the anterior channel and on the spire. Top fawn or olive-green with round white spots and a prominent mantle line.

2.5 cm. Not uncommon at some localities. Central Indo-West Pacific; North-West Cape, W.A. to central N.S.W. The eastern Australian form is sometimes known as *C. l. nashi* IREDALE, 1931. The illustrated shell is from Barrow Is., W.A.

7. *Cypraea cylindrica* BORN, 1778

Cylindrical with a rather rostrate anterior end, columellar teeth numerous, fine and crowded except anteriorly where they are stronger. Base and sides yellow-white, top green-blue with fine diffuse brown freckles and a dark brown irregular central blotch. There are few spots on the sides but prominent dark brown blotches on each side of anterior and posterior ends are a conspicuous feature.

3.5 cm. Common. Central Indo-West Pacific and Micronesia; Shark Bay, W.A. to central Qld. The Qld. form is typical but the western Australian shells (7) are said to show subspecific differences and have been named *C. c. sowerbyana* SCHILDER, 1932.

8, 8a. *Cypraea miliaris* GMELIN, 1791

Solid, broad and strongly pyriform. Teeth strong and, on the columellar side, long. Base and sides porcelaneous white and unspotted, top fawn or olive-green with white spots and a distinct mantle line. Series of pits present along the upper edge of the calloused margins.

5 cm. Uncommon. Indo-West Pacific; Shark Bay, W.A. to central N.S.W. Two subspecies have been recognized in Australia, *C. m. metavona* IREDALE, 1935 in the east and *C. m. diversa* KENYON, 1902 in the west. (8) is a specimen from the Dampier Archipelago, W.A., (8a) is an albinistic form from Fiji which is often recognized as a separate species (or subspecies) named *C. eburnea* BARNES, 1824. Similar pure white specimens are known from Qld. and N.S.W. and this form was named *C. eburnea mara* by Iredale (1931).

9, 9a. *Cypraea helvola* LINNAEUS, 1758

Solid, broad and ovate. Teeth coarse, those on the outer lip extend well across the base, those on the columellar side short except posteriorly where they reach the margin. Sides pitted. Base orange-brown, sides dark brown, top pale brown profusely spotted with brown and white, ends mauve. Mantle line present.

2.5 cm. Common. Indo-West Pacific; Cape Leeuwin, W.A. to central N.S.W. The western and eastern Australian subspecies are *C. h. citrinicolor* IREDALE, 1935 (9, 9a), and *C. h. callista* SHAW, 1909 respectively.

10, 10a. *Cypraea cribraria* LINNAEUS, 1758

Solid, pyriform, teeth strong. Base and sides white, top brown with numerous round white spots.

3.5 cm. Uncommon. Indo-West Pacific; Cape Naturaliste, W.A. to southern Qld. The western Australian subspecies is *C. c. fallax* SMITH, 1881 (10). *C. melwardi* IREDALE, 1930 from Qld. is believed to be an albinistic form of *C. cribraria* (10a).

11, 11a. *Cypraea chinensis* GMELIN, 1791

Solid, ovate-pyriform or elongately subcylindrical. Sides calloused, aperture rather wide. Outer lip teeth strong and extend half way to margin, columellar teeth finer, more numerous and extend well into aperture, Base beige or fawn, with rich orange between the teeth, sides spotted with violet, top blue-grey with orange-brown reticulations.

5 cm. Uncommon. Indo-West Pacific; North-West Cape, W.A. to central N.S.W. The N.S.W. form has been named *C. c. sydneyensis*, SCHILDER & SCHILDER, 1938, and the western Australian form is *C. c. whitworthi* CATE, 1964. (11) is a specimen from North-West Cape; (11a) is from the Philippines.

12. *Cypraea cernica* SOWERBY, 1870

Solid, broad and ovate, with coarse teeth, and pitted sides. Base white, sides white with brown spots, top pale yellow-brown or fawn with white spots and a prominent mantle line.

3.5 cm. Uncommon. Indo-West Pacific; Fremantle, W.A. to central N.S.W. The western subspecies is *C. c. viridicolor* CATE, 1962 and the eastern one is *C. c. tomlini* SCHILDER, 1930 (= *percomis* IREDALE, 1931 = *prodiga* IREDALE, 1939). The illustrated specimen is from North-West Cape, W.A.

Plate 32. Cypraeidae (1%/10 × natural size)

Cypraeidae (Cont.)

1, 1a. *Cypraea subviridis* REEVE, 1835
Pyriform and rather thin, with a sinuous aperture and numerous weak teeth. Base and sides white, sometimes pinkish, top green with crowded minute brown freckles and faint bands.

4 cm. New Caledonia and northern Australia from Rottnest Is., W.A. to central N.S.W. 3 rather distinctive subspecies are recognized in this species. W.A. shells have a large conspicuous brown dorsal blotch and are known as *C. s. dorsalis* SCHILDER & SCHILDER, 1938 (1, 1a). The typical form from Qld., *C. s. subviridis*, lacks the blotch. Iredale (1931) gave the name *C. s. vaticina* to N.S.W. shells which have coarser anterior teeth.

2. *Cypraea hungerfordi coucomi* SCHILDER, 1964
Thin, pyriform and slender. Right hand margin projecting beyond the left one at the posterior end. Outer lip teeth short and fairly coarse, columellar teeth short and fine, especially posteriorly. Base and sides pale yellow-orange, with a few large brown spots on the sides, top pale orange to fawn with rust-coloured flecks and 3 transverse bands or zones of darker blotches.

4.3 cm. Uncommon. Trawled in 60 to 100 fathoms off Cape Moreton Qld. Schilder originally described this as a full species but later reduced it to the status of a subspecies of the Japanese *C. hungerfordi* SOWERBY, 1888 when additional Qld. specimens showed closer resemblance than he at first thought existed.

3. *Cypraea queenslandica* SCHILDER, 1966
Thin, pyriform, swollen in the middle and rostrate and sharply margined at the sides. Outer lip teeth fine and extend slightly over the base; columellar teeth finer still and more numerous, especially near the centre. Base and sides white to pale lemon, with pale brown spots on the sides. Top whitish with a large irregular central chestnut-brown blotch, and crowded yellow-brown spots which form an irregular net pattern.

5.7 cm. Only 1 specimen is known. The figured specimen is the holotype, now in the collection of Greenacres Shell Museum, Dunsborough, W.A. Trawled at 70 fathoms off Cape Moreton, Qld.

4, 4a. *Cypraea stolida* LINNAEUS, 1758
Cylindrical to oblong-ovate, with slightly beaked ends, strong teeth extending over the base, and a depressed spire. Base yellow-white, sides with indistinct brown spots. Light orange-brown blotches present on each side of the anterior and posterior ends, and 2 blotches of the same colour on each side. Top blue-grey with indistinct brown spots, and a large irregular brown central blotch which is usually partially connected to the side blotches.

3.5 cm. Uncommon. Indo-West Pacific; Geraldton, W.A. to Qld. Both figured shells are from Dampier Archipelago, W.A. (Compare with *C. brevidentata*).

Plate 33. *Cypraea stolida* LINNAEUS, 1758, with eggs. Humpy Isle, Qld., at a depth of 20 feet. (*Photo*—Neville Coleman).

5. *Cypraea caputserpentis* LINNAEUS, 1758
Solid, depressed, with strong teeth, margins thickened and expanded to form a broad flat base. Base and sides dark chocolate brown, top brown with numerous white spots. Juvenile shells have a prominent central band.

3.5 cm. Very common. Indo-West Pacific; found as far south as Albany on the south coast of W.A., and southern N.S.W. on the east coast. Southern shells on both sides of Australia have a rather juvenile appearance, lacking strongly thickened margins. Seven subspecies have been named in this widespread species, including 4 in Australian waters, but the supposed differences between them are very minor. We refer the reader to the works of Schilder & Schilder (1938-39) and Allan (1956) for details.

6, 6a. *Cypraea walkeri* SOWERBY, 1832
Pyriform with long teeth on both sides of the aperture. Base pale orange, conspicuously stained with lilac between the teeth, sides pale orange with prominent dark brown spots. Large brown blotches present on each side of the anterior end and another on the depressed spire. Top greenish with numerous minute brown freckles, with a broad almost continuous brown band centrally, separated in front and behind by thin white lines from bands of squarish spots.

3.5 cm. Not common. North Indian Ocean and central Indo-West Pacific; Broome, W.A. to central Qld. The Australian subspecies is *C. w. continens* IREDALE, 1935.

7, 7a. *Cypraea poraria* LINNAEUS, 1758
Solid, ovate or deltoidal, margins calloused and pitted anteriorly. Teeth strong and moderately short except at the anterior and posterior ends of the columella where they extend well over the base. Base and sides deep lavender, teeth and interstices white, top brown with white spots ringed with darker brown or violet. Thin mantle line present.

2.5 cm. Uncommon. Indo-West Pacific; found in the North-West Cape area of W.A. and in Qld. and northern N.S.W. Western Australian shells belong to the Indian Ocean subspecies *C. p. poraria* but eastern Australian shells are known as *C. p. scarabaeus* BORY, 1827.

8. *Cypraea brevidentata* SOWERBY, 1870
Ovate to pyriform, with beaked ends, a depressed spire, and strong teeth which do not extend far across the base. Base white, sides with or without indistinct brown spots, 2 small brown blotches present on each side, and 1 on each side of the anterior and posterior ends. Top blue-grey without spots, but with a large sharply margined brown central blotch.

2.5 cm. Uncommon. Northern W.A. between North-West Cape and Broome. This species was once considered a subspecies of *C. stolida* but the shells of the 2 species are distinct and they live together in W.A.

9, 9a. *Cypraea xanthodon* SOWERBY, 1832
Solid and pyriformly ovate, with short fairly strong and well spaced teeth. Base orange-brown (darker on the teeth); sides with large dark brown spots. There are prominent brown blotches on each side of the anterior end and on the spire. Top blue-green with numerous small brown spots, and sometimes 3 dark bands and a large central brown blotch.

3 cm. Common. Qld. and northern N.S.W.

10. *Cypraea langfordi moretonensis* SCHILDER, 1965
Rather solid, pyriform, anterior end slightly rostrate and margin-ated. Strong teeth border the narrow and slightly curved aperture. Base orange, sides red-orange and unspotted with irregular rust-coloured flecks on the white top, and a darker rust-coloured blotch on the right side of the spire.

6.5 cm. Uncommon. Trawled in 60 to 100 fathoms off Cape Moreton, Qld. These Qld. shells are very like the rare Japanese *C. langfordi* KURODA, 1938 and Schilder named them a subspecies mainly because of the distance separating the Japanese and Qld. populations.

11, 11a. *Cypraea pyriformis* GRAY, 1824
Pyriform, with numerous fine teeth which are conspicuously red coloured on the columellar side. Base and sides white with dark brown spots on the sides and on either side of the anterior end. A dark brown blotch present on the rather depressed spire. Top green with numerous minute brown freckles and a broad central interrupted band.

3 cm. Not common. Central Indo-West Pacific; Kimberley coast, northern W.A. to central Qld.

Plate 34. Cypraeidae (1%₁₀ × natural size)

Cypraeidae (Cont.)

1, 1a. *Cypraea lutea* GMELIN, 1791
Rather thin, ovate to pyriform, with short but well developed teeth. Base and sides rich orange-brown with distinct dark brown spots on the sides. Top deep orange or fawn with 2 thin white spiral lines, and often with scattered brown spots.
2 cm. Uncommon. Indo-West Pacific; North-West Cape to the Kimberleys, W.A. Western Australian shells have been named *C. l. bizonata* IREDALE, 1935; they were believed to lack the scattered dorsal brown spots characteristic of the typical form from the Malaysian-Philippine region, but this difference is not consistent (e.g. (1) from Broome, W.A.). The Qld. cowry known as *C. humphreysii* GRAY, 1825 (13) is often considered to be merely another subspecies of *C. lutea.*

2. *Cypraea moneta* LINNAEUS, 1758
Broad, solid, and depressed, with irregular knobby margins, short and strong teeth, and a narrow aperture. Base and sides white or yellow, without spots, top yellow, orange or green, with 3 faint darker bands and sometimes a faint yellow encircling line. Variable.
3 cm. Very common. Indo-West Pacific; Houtman Abrolhos, W.A. to northern Qld. The western and eastern Australian subspecies are *C. m. rhomboides* SCHILDER & SCHILDER, 1933, and *C. m. barthelemyi* BERNARDI, 1861, respectively.

3. *Cypraea asellus* LINNAEUS, 1758
Solid and pyriformly ovate. Teeth short except on the posterior end of the columella. Porcellaneous white except for 3 broad bands of black or dark brown across the top.
2 cm. Uncommon. Indo-West Pacific; North-West Cape. W.A. to southern N.S.W. Western and northern Australian shells are typical but the N.S.W. shells are said to be more dilated at the ends and to have shorter posterior columellar teeth. Schilder (1930) proposed the subspecies name *latefasciata* for N.S.W. specimens.

4. *Cypraea pallidula* GASKOIN, 1849
Solid, subcylindrical, with strong teeth which extend well across the base on the columellar side. Base and sides white, sides sometimes spotted with brown, top green-grey with numerous minute brown freckles and 4 distinct interrupted bands of darker grey.
3 cm. Uncommon in central Indo-West Pacific; Cape Naturaliste, W.A. to central Qld..Qld. and W.A. shells have been given distinct subspecies status and named *C. p. rhinoceros* SOWERBY, 1865, and *C. p. simulans* SCHILDER & SCHILDER, 1940 respectively. The figured shell is from the Dampier Archipelago, W.A.

5. *Cypraea annulus* LINNAEUS, 1758
Solid, broad and flat-based, with strong teeth which extend well across the base on both sides. Cream-white, yellowish or greyish, sides lack spots, top bluish-white with a conspicuous yellow encircling ring.
3 cm. Very common. Indo-West Pacific; Rottnest Is., W.A. to southern N.S.W. The western Australian form is typical but the eastern Australian one belongs to the subspecies *C. a. noumeensis* MARIE, 1869.

6, 6a. *Cypraea felina* GMELIN, 1791
Flat based, small, solid, and depressed. Teeth numerous but weak on the columellar side. Base white, sides white with large irregular brown spots. Anterior and posterior ends with large conspicuous brown blotches. Top blue-grey with 4 bands of large squarish brown spots.
2.5 cm. Uncommon. Indo-West Pacific; Qld. and northern N.S.W. There is a single record from W.A. at North-West Cape. The Australian subspecies is *C. f. melvilli* HIDALGO, 1906.

7. *Cypraea quadrimaculata* GRAY, 1824
Solid, elongate-ovate, spire slightly umbilicate. On the columellar side the teeth extend across the base almost to the margins. Base and sides creamy-white and unspotted. Top green-blue with profuse minute pale brown freckles. Conspicuous dark brown spots present on each side of the anterior and posterior ends.
3.5 cm. Uncommon. Central Indo-West Pacific and Micronesia; Kimberley coast, W.A. to central Qld. The Australian subspecies is *C. q. thielei* SCHILDER & SCHILDER, 1938.

8. *Cypraea saulae* GASKOIN, 1843
(see also colour plate of live animal, page 42, plate 19)
Pyriform, slightly rostrate at the anterior end, with an umbilicate spire and short but strong teeth. Creamy-white or pale fawn,

slightly darker at the ends. Small but conspicuous dark brown spots present on the sides, top with small scattered pale brown spots, and a prominent irregular brown central blotch.
3 cm. Very uncommon. Central Indo-West Pacific; Dampier Archipelago, W.A. to central Qld. Qld. specimens are known as *C. s. nugata* IREDALE, 1935 and western Australian shells as *C. s. crakei* CATE, 1968. The illustrated shell is from the Dampier Archipelago, W.A.

9. *Cypraea hammondae* IREDALE, 1939
Very like *C. fimbriata* but more pyriform or ovate. Top with few freckles but numerous small orange-brown spots.
2 cm. Uncommon. Central Indo-West Pacific; Houtman Abrolhos, W.A. to northern N.S.W. The type specimen of the species came from N.S.W. and similar shells are recorded from Qld. and Arnhem Land. Western Australian shells (e.g. 9) are more slender, have fewer teeth, a greyish top (instead of whitish) and a pale brown base (instead of white). They have been recognized as a subspecies, viz. *C. h. dampierensis* SCHILDER & CERNOHORSKY, 1965.

10. *Cypraea fimbriata* GMELIN, 1791
Small, subcylindrical, with slightly beaked ends and fine and numerous teeth. Base white, sides with prominent dark brown spots, top pale orange-brown or grey with numerous minute brown freckles and often 3 bands of heavier brown blotches (the central band is often the most prominent). Conspicuous lilac spots on each side of the anterior and posterior ends and a brown spire blotch are characteristic.
2 cm. Uncommon. Indo-West Pacific; Houtman Abrolhos, W.A. to central N.S.W. The Australian form is typical although Iredale (1939) gave the subspecific name *blandita* to N.S.W. specimens.

11. *Cypraea microdon* GRAY, 1828
Tiny, slender, pyriform, with the left side of the posterior canal long, thick and projecting. Teeth small and numerous. Base and sides white, left side with a few small brown spots. Top white or very pale orange-brown, with crowded minute brown freckles and 2 pale bands bordered on each side by faint brown spots in some shells. Ends with lilac spots like those of *C. fimbriata* and *C. hammondae.*
1 cm. Uncommon. Indo-West Pacific and Micronesia; Qld.

12, 12a. *Cypraea gracilis* GASKOIN, 1849
Small, rather solid, pyriform to ovate, with fine and numerous teeth. Base and sides white, sides with conspicuous brown spots, top blue-grey with numerous small brown freckles and some larger spots, and usually a large central brown blotch. Faint lilac spots present at the anterior and posterior ends.
2.5 cm. Common in the Indo-West Pacific; Cape Naturaliste, W.A. to northern N.S.W. Iredale (1939) named the western shell *hilda*, thus separating it from the eastern Australian *macula* ANGAS, 1867 (12,12a) which he regarded as a distinct species. However, both forms are worthy of no more than subspecific status.

13. *Cypraea humphreysii* GRAY, 1825
Like *C. lutea* but the top blue-white with numerous crowded brown spots, and the white spiral lines rather broad and indistinct.
1.5 cm. Uncommon. South-Western Pacific; Qld. to northern N.S.W. Compare with *C. lutea.*

14. *Cypraea ziczac* LINNAEUS, 1758
Light, pyriform, with small, numerous and short teeth. Base orange-brown with dark brown spots which extend to the sides, top fawn or light brown, with bands of brown spots at each end and 3 transverse white bands containing V-shaped light brown lines.
2.5 cm. Uncommon. Indo-West Pacific; Shark Bay, W.A. to central N.S.W. The eastern Australian form is known as *C. z. vittata* DESHAYES, 1831, while western Australian shells are said to be *C. ziczac* in the strict sense.

15. *Cypraea clandestina* LINNAEUS, 1767
Small, solid and subpyriform, with strong and rather long teeth. Base and sides creamy-white, top ivory-white with fine faint pale brown zigzag lines reaching from one side to the other.
2.5 cm. Common. Indo-West Pacific; Houtman Abrolhos, W.A. to central N.S.W. The western Australian shells are typical of the nominate subspecies but Iredale (1939) proposed the specific names *whitleyi* and *extrema* for Qld. and N.S.W. forms respectively.

Plate 35. Cypraeidae (1⁹⁄₁₀× natural size)

Cypraeidae (Cont.)

Plate 36. *Cypraea nucleus* LINNAEUS, 1758. Arafura Sea.
(Photo—Barry Wilson).

1. *Cypraea limacina* LAMARCK, 1810
Not very solid, elongately pyriform, margins calloused and pitted.
Teeth do not reach the margins. Base white, teeth and ends orange-
brown, top bluish or purple-grey with white spots which may be
nodular on the sides. A weak mantle line present.
3.5 cm. Moderately common. Indo-West Pacific; Cape Naturaliste, W.A. to N.S.W. *C. l. facifer* IREDALE, 1935 is the subspecific
name usually applied to the Australian form.

2, 2a *Cypraea staphylaea* LINNAEUS, 1758
(see also colour plate of live animal, page 45, plate 21)
Solid, ovate, margins calloused and pitted, teeth extend across
base to the margins. Base white, teeth and ends orange-brown,
top yellow-brown or grey with nodular white spots. A mantle
line present.
3 cm. Uncommon. Indo-West Pacific; Houtman Abrolhos, W.A.
to N.S.W. The 2 recognized Australian subspecies are *C. s.
staphylaea* in the west and *C. s. descripta* IREDALE, 1935 in the east.

3, 3a. *Cypraea nucleus* LINNAEUS, 1758
Solid, broadly ovate, depressed, margins calloused. Upper surface
with coarse nodules and a deep, winding sulcus or mantle line.
Teeth strong, bifurcate and extend across base and up on to the
sides. Base and sides white, pale fawn or orange, teeth edged with
orange-brown, top yellow-brown, nodules ringed with darker
red-brown.
·3 cm. Uncommon. Indo-West Pacific; North-West Cape and
Dampier Archipelago areas, W.A.

4, 4a. *Cypraea hirundo* LINNAEUS, 1758
Solid, ovate to subcylindrical. Teeth fine and extend across the base
almost to the margins. Base white, top pale blue with 3 poorly
defined blue zones separated by white bands, and fine brown spots
most numerous on the sides. A prominent brown dorsal blotch
present and conspicuous dark brown blotches at the ends.
2 cm. Not uncommon at some localities. Indo-West Pacific;
Shark Bay, W.A. to northern N.S.W. The Australian subspecies is
C. h. cameroni IREDALE, 1939.

5. *Cypraea kieneri* HIDALGO, 1906
Small, solid, ovate, rather slender. Teeth rather coarse, outer lip
teeth short, columellar teeth short anteriorly but extend well across
the base posteriorly. Base white, top with 3 well defined but
irregular blue zones separated by rather broad white wavy bands,
anterior blue zone enclosing a paler patch. No dorsal blotch present
but sometimes there are squarish brown spots in the central zone.
Ends with conspicuous dark brown blotches.
2 cm. Uncommon. Indo-West Pacific; North-West Cape, W.A.
to central N.S.W.

Plate 37. *Cypraea (Notocypraea) piperita* GRAY, 1825.
Cape Naturaliste, W.A.   *(Photo*—Barry Wilson).

6. *Cypraea ursellus* GMELIN, 1791
Small, rather solid, pyriform. Teeth fine and numerous and extend
across the base almost to the margins. Base white, top with 3 well
defined blue-grey zones of irregular outline separated by white
bands, and scattered fine brown spots. Ends with conspicuous dark
brown spots.
1 cm. Uncommon. Indo-West Pacific; Broome, W.A. to northern
N.S.W. Iredale (1939) named the N.S.W. form *C. u. marcia*.

7, 7a. *Cypraea punctata* LINNAEUS, 1771
(see also colour plate of live animal, page 45, plate 22)
Oblong-ovate to pyriform, right margin thickened and slightly
angulate, left margin rounded, spire depressed, teeth short and fine.
Milky-white with medium sized brown spots on the top and margins.
2.5 cm. Uncommon. Indo-West Pacific; eastern Qld. SCHILDER &
SCHILDER (1938) named the Qld. and south-western Pacific form
C. p. iredalei.

8. *Cypraea (Notocypraea) molleri* IREDALE, 1931
Ovate, rather thin and tumid. Outer lip teeth extend part way across
the base, but those on the columellar side reach only just beyond
the aperture. Fossula wide, deep and long, fossular teeth divided.
Top pale flesh or white, sometimes with 4 pale indistinct interrupted
transverse bands, right side sparsely spotted, left side very sparsely
spotted or unspotted.
3 cm. Uncommon. Green Cape, N.S.W. to Lakes Entrance, Vic.
Schilder (1964) considers this to be a subspecies of *C. angustata*.

9, 9a. *Cypraea (Notocypraea) angustata* GMELIN, 1791
Swollen, humped and broadly ovate. Outer lip teeth extend part
way across the base and finer than the columellar teeth. Fossula
shallow, fossular teeth often divided into upper and lower parts.
Base white, sides heavily spotted with brown, conspicuous dark
brown terminal blotches present. Top dark brown or grey-brown,
and usually lacks transverse bands or spots.
3 cm. Common. Southern N.S.W. to S.A. and Tas. *C. verconis*
COTTON & GODFREY, 1932 is a synonym.

10, 10a. *Cypraea (Notocypraea) declivis* SOWERBY, 1870
Swollen, humped and broadly ovate. Outer lip and columellar
teeth fine, fossula shallow, fossular teeth continuous and not
deeply divided. Base white, sides with fine brown spots, top pale
milky-fawn or sepia densely flecked with diffuse small pale brown
specks and with conspicuous dark brown blotches at the ends.
2.5 cm. Fairly common. Vic., Tas. and S.A. Iredale's (1935)
subspecies *C. d. occidentalis* from southern W.A. is probably a
form of *C. piperita*.

11, 11a. *Cypraea (Notocypraea) pulicaria* REEVE, 1846
Cylindrical and elongate. Teeth on both sides of aperture very fine.
Fossula deeply concave with a projecting inner edge. Base, sides and
top rose-pink or orange, top conspicuously crossed by 4 darker
interrupted transverse bands. Brown spots present on the sides,
terminal blotches small or obsolete.
2 cm. Common. Southern W.A. between Cape Leeuwin and
Rottnest Is.

12, 12a, 12b. *Cypraea (Notocypraea) comptoni* GRAY, 1847
Sub-pyriform, rather slender, with fine short teeth. Fossula shallow,
fossular teeth often divided into upper and lower parts. Base white,
sides with many fine brown spots, top red-brown. Top usually with
4 darker transverse bands, the central pair more or less continuous
and more conspicuous. Brown blotches present at the ends.
2.5 cm. Common. Southern N.S.W. to Cape Leeuwin, W.A.
Colour is variable and white forms (e.g. *Notocypraea casta*
SCHILDER & SUMMERS, 1963 (12b)) and unbanded forms occur,
especially in S.A.

13, 13a, 13b. *Cypraea (Notocypraea) piperita* GRAY, 1825
Pyriform, right side marginated. Outer lip teeth fine and short,
columellar teeth fine. Fossula concave, fossular teeth well formed
and divided into upper and lower parts. Base cream or fawn, sides
heavily spotted, top cream or pale orange-brown with 4 wide
interrupted bands (anterior and posterior bands sometimes
obsolete) and usually with red-brown spots and diffuse flecks. There
is sometimes a reticulate pattern of lines. Mantle line often present
but terminal blotches obsolete.
2.5 cm. Common. Southern N.S.W. to Cape Naturaliste, W.A.
This is a very variable species. The name *C. reticulifera* SCHILDER,
1924 refers to an unspotted form of *C. piperita* with narrow interrupted transverse bands (13b). *C. bicolor* GASKOIN, 1848 is also
referable to this species.

Plate 38. Cypraeidae (1$\frac{8}{10}$× natural size)

egg and spindle cowries (FAMILY OVULIDAE)

THE OVULIDS OR EGG AND SPINDLE COWRIES are a diverse group of molluscs related to the true cowries. Most of the species have very brightly coloured and patterned mantles. They all feed on the polyps of some form of coelenterate and, in order to find living specimens, one must first identify the host coelenterate. For example, the large white egg cowry, *Ovula ovum,* is found associated with large fleshy soft corals; species of the genera *Volva* and *Neosimnia* live on the fronds of gorgonians or fan-corals. Often the host coelenterate occurs in several different colours (some species of gorgonian may be orange, red or yellow) and in such cases the ovulid shells may take on the colour of the host. Nearly 100 living species are known and there are many genera. Only a few common species are illustrated.

Selected references:

SCHILDER, F. A. (1932). The living species of Amphiperatinae. Proc. malac. Soc. London, **20** : 40-64, pls 3-5.

Plate 39. *Volva volva* LINNAEUS, 1758, Qld. (*Photo*—Don Byrne).

Plate 40. *Primovula (Diminovula) punctata* DUCLOS, 1831. A common species from Qld. and the central Indo-West Pacific. (*Photo*—Don Byrne).

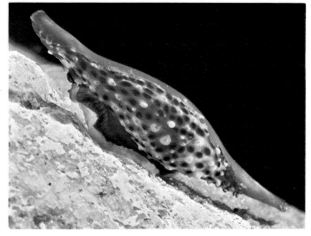

Plate 41. *Volva nectarea* IREDALE, 1930. A species from N.S.W. (*Photo*—Keith Gillett).

Plate 42. *Ovula ovum* LINNAEUS, 1758. At a depth of 36 feet on the alcyonarian, *Sarcophyton trocheliophorum* MARENZELLER, 1886. (*Photo*—Neville Coleman).

Cypraeidae (Cont.)

1, 1a. *Cypraea globulus* LINNAEUS, 1758
Globular, with beaked projecting ends, rounded sides and base, and teeth which extend across the base but do not reach the margins. Upper surface smooth, pale to moderately dark orange-brown and heavily spotted with darker brown, 2 pairs of dark brown blotches conspicuous on the base. There is no mantle sulcus (i.e. the mantle line is weak, not incised).
    2.5 cm. Uncommon. Indo-West Pacific; North-West Cape, W.A. to Qld. This species, and its relatives *C. cicercula* and *C. bistrinotata*, belong in a group of small, curious cowries usually grouped in the genus or subgenus *Pustularia*.

2. *Cypraea bistrinotata* SCHILDER & SCHILDER, 1937
Globular with beaked, projecting ends, rounded sides and base and teeth which extend across the base to the margins. Upper surface granulose and light yellow-brown. 2 pairs of brown blotches present on the base, and 3 pairs of brown blotches and many small spots on the top. A deep mantle sulcus (i.e. an incised mantle line) passes between the dorsal spots.
    2 cm. Uncommon. Indo-West Pacific; Qld.

3. *Cypraea cicercula* LINNAEUS, 1758
Globular with beaked, projecting ends, rounded sides and base and teeth which extend across the base to the margins. The upper surface granulose, pale yellow, with faint brown spots but no blotches. A mantle sulcus present.
    2.5 cm. Indo-West Pacific; North-West Cape, W.A. to Qld.

(FAMILY OVULIDAE)

4, 4a. *Volva volva* LINNAEUS, 1758
(see also colour plate of live animal, page 60, plate 39)
Thin, swollen centrally, with the ends drawn out to form long, curved, slender canals. Outer lip rounded and calloused, teeth obsolete. Columella smooth, fossula obsolete. Fine spiral striae present on the ends of the canals, body whorl smooth and grey-cream, sometimes tinged with pink. Lip white or pink.
    10 cm. Moderately common. Indo-West Pacific; North-West Cape, W.A. to N.S.W.

5, 5a. *Calpurnus verrucosus* LINNAEUS, 1758
(see also colour plate of live animal, page 19, plate 6)
Solid, ovate, broad and flat based, with a triangular keel across the back. Coarse teeth present on outer lip, columella smooth. Top with minute transverse striae and button-like tubercles at the ends. White, pink at the ends.
    3.5 cm. Common. Indo-West Pacific; Qld.

6. *Ovula ovum* LINNAEUS, 1758
(see also colour plate of live animal, page 61, plate 42)
Solid, ovate and inflated with rostrate ends. Outer teeth weak, and irregular columellar teeth. Exterior porcellaneous white, interior orange-brown.
    12 cm. Common. Indo-West Pacific; Houtman Abrolhos, W.A. to northern N.S.W. Once known under the generic name *Amphiperas*.

7. *Ovula costellata* LAMARCK, 1810
Solid, pyriform, inflated. Top obsoletely carinated, left side deeply notched below the spire, anterior extremity broad and flattened. Outer lip rounded and thickened with weak irregular teeth, columella smooth. Exterior porcellaneous white with fine transverse striae, interior pink.
    5 cm. Uncommon. Indo-West Pacific; Qld.

8. *Volva philippinarum* SOWERBY, 1848
Thin, long and slender, with the ends drawn out and tapered to form straight canals. Canals finely spirally striate but the body of the shell smooth. Outer lip thickened and rounded, outer lip and columella without teeth, fossula lacking. Colour of the whorls varies depending on the colour of the host gorgonian, may be pink, red, orange or yellow, outer lip usually white.
    3.5 cm. Uncommon. Central Indo-Pacific; W.A. at least as far south as Fremantle. Specimens like this from the south coast of W.A. may be a separate species.

9. *Neosimnia depressa* SOWERBY, 1875
Thin, narrow and lanceolate, with tapered ends forming fairly short canals. Aperture narrow, widening anteriorly, outer lip turned inwards and flattened, edentulous, flared out anteriorly. Columella smooth, fossula sharp-edged but narrow. Apricot to rose-pink with orange-brown tips.
    3 cm. Common. Dampier Archipelago to Broome, W.A. The generic characters separating *Neosimnia* from *Volva* are not consistent and are difficult to apply.

10. *Prionovolva cavanaghi* IREDALE, 1931
(see also colour plate of live animal, page 5, plate 1)
Solid, ovate to globular with a rounded outer lip which lacks teeth or bears only weak folds. Columella smooth, fossula concave; anterior terminal ridge on the columellar side sharp and strong, funiculum weak, no basal carina on the columellar side. Pink with white ends.
    1.5 cm. Central Indo-West Pacific; North-West Cape, W.A. to N.S.W.

trivias (FAMILY TRIVIIDAE)

THE TRIVIAS ARE SMALL, globular, cowry-like shells with the apertural teeth continued over the sides and top. They are closely related to the true cowries. There are several genera and many living species but their taxonomy is confusing. Only 2 common Australian species are illustrated here.

Plate 43. *Trivirostra oryza* LAMARCK, 1810, Qld.
                    (*Photo*—Don Byrne).

11. *Trivirostra oryza* LAMARCK, 1810
Small and ovate to globular. Ends slightly drawn out, spire not visible. Dorsal ribs fine but sharply defined and terminate at a deep, longitudinal, median, dorsal groove. White.
    0.8 cm. Common. Indo-West Pacific. Shark Bay, W.A. to southern Qld.

12. *Ellatrivia merces* IREDALE, 1924
Small and ovate, with the spire evident. Ribs strong on the sides but become obsolete dorsally so that the dorsum is smooth at the centre. White with a brown central blotch or blotches on top, and pink or red-brown ends.
    1.3 cm. Very common. N.S.W. to Geraldton, W.A.

Plate 44. Cypraeidae, Ovulidae and Triviidae (1⁹⁄₁₀ × natural size) ⤷

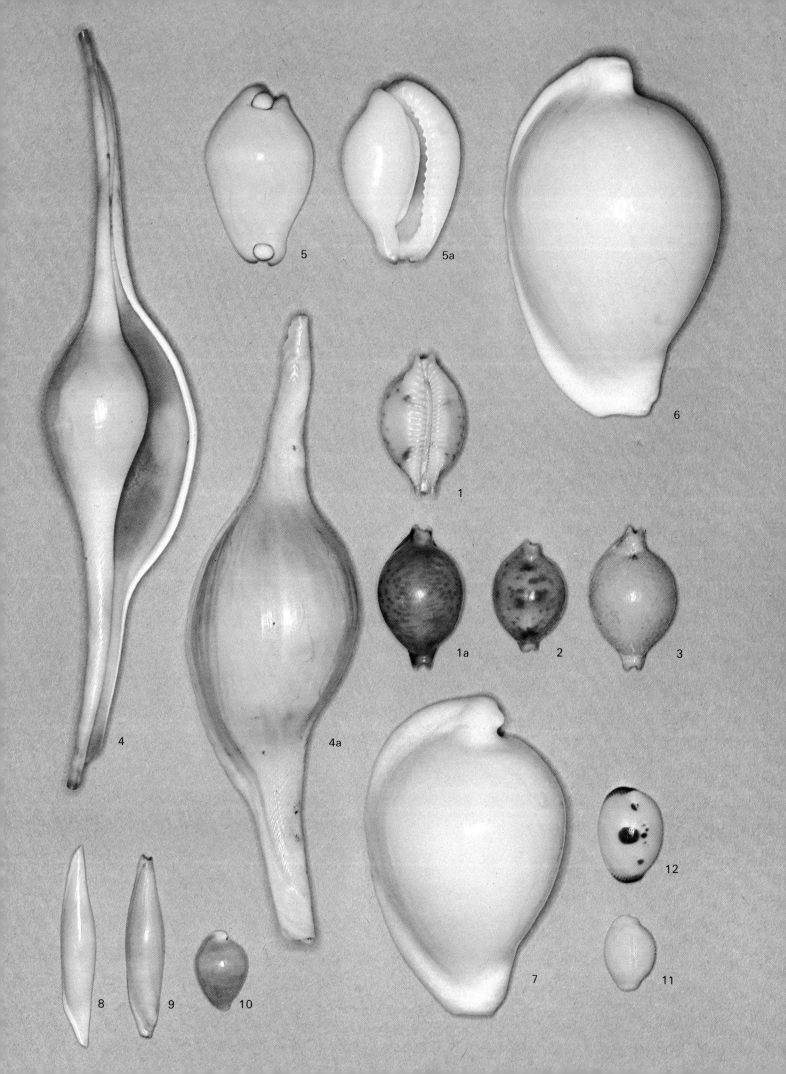

moon shells and naticas (FAMILY NATICIDAE)

NATICAS ARE AMONG THE MORE CONSPICUOUS MOLLUSCS of intertidal sand flats. They prey upon other molluscs, especially bivalves, and like most predators are quite active animals. The foot is greatly expanded with lobes that enclose or nearly enclose the shell, and the animal ploughs along just below the surface of the sand leaving behind a wide trail. Naticas are able to drill neat round holes through the shells of their prey by means of circular movements of the specially adapted radula aided by softening acid secretions. The proboscis is inserted through the hole and the body of the prey is consumed. Females lay gelatinous egg-masses of 2 quite different kinds. At one time it was thought that all species laid sand-impregnated "egg-collars" but a few years ago the late Miss Fay Murray of Melbourne showed that the common southern moon shells *Polinices conicus*, *P. sordidus*, and *P. incei* lay clear gelatinous sausage-shaped egg-masses.

There are also 2 kinds of operculum in this family, thin and horny in the subfamilies Mammillinae and Sininae, thick and calcareous in the subfamily Naticinae.

1, 1a, 1b. *Polinices sordidus* SWAINSON, 1821
Conically ovate, spire moderately high. Umbilicus deep, partly closed by callus. Exterior blue-grey with an orange spiral band at the sutures, interior chocolate brown, umbilicus and callus orange or red-brown. Operculum horny.
 5 cm. Common. Eastern and south-eastern Australia from northern Qld. to S.A. Distinguished from its near relative *P. conicus* LAMARCK, 1822 (not illustrated) by its more globose shape and darker grey external colour.

2, 2a. *Polinices mellosus* HEDLEY, 1924
Solid; conically ovate; polished, porcellaneous; spire moderately high. Umbilicus completely filled by a large callus. Exterior pale yellow, aperture and umbilical callus white. Operculum horny.
 4.5 cm. Common N.T. to eastern Qld. The yellow colour and less elongate shape readily distinguished this species from *P. pyriformis*.

3, 3a. *Polinices pyriformis* RECLUZ, 1844
Solid, conically ovate, porcellaneous; spire high. Umbilicus usually completely filled by a large callus. Exterior and interior polished, white. Operculum horny.
 5 cm. Common. Indo-West Pacific; Fremantle, W.A. to eastern Qld.

4, 4a, 4b. *Polinices incei* PHILIPPI, 1851
Rounded, very depressed, spire low. Umbilicus filled by a button-like callus leaving only a deep semi-circular groove. Upper surface pale yellow, white, grey, brown or purple, lower surface usually white. Operculum horny.
 3 cm. Common. Eastern and south-eastern Australia (excluding Tas.) from Qld. to S.A.

5, 5a. *Mamillaria powisiana* RECLUZ, 1844
Solid, porcellaneous, conically ovate, spire of medium height. Umbilicus wide, deep and spiral, partly closed by a heavy callus. White with a wide central orange-brown spiral band on body whorl. Operculum horny.
 6 cm. Moderately common. Indo-West Pacific; Shark Bay, W.A. to eastern Qld.

6, 6a. *Mammilla opaca* RECLUZ, 1851
Thin, elongately ovate, polished, spire of medium height. Umbilicus small, almost closed by a reflected callus from the columella. Exterior off-white with a white spiral band at the sutures and 2 or 3 obscure spiral bands of flesh-brown on the body whorl, columellar callus and umbilicus dark chocolate brown. Operculum horny.
 4.5 cm. Common. Indo-West Pacific; Pt. Cloates, W.A. to eastern Qld.

7. *Mammilla sebae* SOULEYET, 1844
Very thin, semi-transparent, ovate, rather flat, spire low. Umbilicus narrow, almost closed by a long narrow reflected columellar callus. Surface dull white with 2 spiral bands of light brown blotches on the body whorl and a spiral row of light brown axial dashes below the suture, columella and callus brown. Operculum horny.
 5 cm. Common. Indo-West Pacific; eastern Qld. The very thin semi-transparent shell and the long narrow columella distinguish this species from *M. simiae*.

8, 8a. *Mammilla simiae* DESHAYES, 1838
Thin, ovate, spire moderately low. Umbilicus small almost closed by a rather broad reflected callus from the columella. Body whorl with very fine spiral striae. Exterior a dull off-white with 3 wide spiral bands of irregular brown axial streaks, columella and callus chocolate brown. Operculum horny.
 5 cm. Common. Indo-West Pacific; Dampier Archipelago, W.A. to N.S.W.

9, 9a. *Naticarius alapapilionis* RÖDING, 1798
Globose, spire of medium height, whorls slightly flattened below the sutures. Aperture wide, outer lip slightly flared. Umbilicus wide and deep, with a thick central funicle forming a callus at the columellar margin. Exterior fawn or blue-grey with 4 white spiral lines which contain brown spots. Operculum calcareous with many deep spiral striae.
 4 cm. Moderately common. Indo-West Pacific; eastern Qld. and N.S.W.

10, 10a. *Natica solida* BLAINVILLE, 1825
Moderately solid, ovate to globose, spire low. Umbilicus narrow, almost closed by a broad reflected columellar callus, funicle weak. Exterior brown with 2 broad lighter spiral bands, 1 at the sutures and the other near the anterior end, columella and umbilical callus glossy rich chocolate brown. Operculum calcareous, smooth except for a few weak spiral striae near the margin of the curved side and small marginal serrations on the straight side.
 3 cm. Common. Central Indo-West Pacific; Dampier Archipelago, W.A. (Qld.?)

11, 11a. *Naticarius oncus* RÖDING, 1798.
Moderately solid, conically ovate, spire of medium height. Umbilicus wide with a broad internal central funicle forming a callus at the columellar margin. Exterior polished white with 2 to 5 spiral rows of red-brown spots. Operculum calcareous, with spiral grooves.
 2.5 cm. Moderately common. Indo-West Pacific; N.T. to eastern Qld. Often known by the name *N. chinensis* LAMARCK, 1822.

12, 12a. *Notocochlis sagittata* MENKE, 1843
Moderately solid, conically ovate to globose, spire of medium height. Umbilicus small with a thick central funicle forming a callus at the columellar margin. Exterior white or pale fawn, with light brown, fine, curved, axial lines, a spiral band of brown patches at the sutures, and 2 central white bands on the body whorl (each of which contain light brown arrow-head marks which point toward the margin). Interior and umbilical callus white. Operculum calcareous with weak marginal spiral grooves.
 1.5 cm. Common. Found on all Australian coasts.

13, 13a. *Sigaretotrema umbilicata* QUOY & GAIMARD, 1833
Thin, ovate to globose, spire low. Umbilicus narrow but deep and rather straight, sometimes partly occluded by a weak columellar callus. Exterior surface with very fine spiral striae and growth striae. Off white or cream with weakly developed axial streaks of pale brown, a white spiral band at the sutures, and a broad white central spiral band on the body whorl which interrupts the axial streaks. Operculum horny.
 2.5 cm. Common. N.S.W. to Fremantle, W.A.

14, 14a. *Ectosinum zonale* QUOY & GAIMARD, 1833
Very flat and ear-shaped with a small depressed spire and wide shallow aperture. Upper surface with crowded wrinkled spiral striae, interior glossy smooth. White except for a purple-brown apex, and sometimes reddish radial streaks. Operculum lacking.
 4 cm. Common. Southern Qld. to Rottnest Is., W.A. The animal is very large with lateral flaps that enclose the shell.

Plate 45. Naticidae ($1\frac{2}{10}$× natural size)

helmet shells (FAMILY CASSIDAE)

HELMET OR CASSID SHELLS are handsome medium sized to large molluscs which are common household ornaments. A comprehensive monograph of the family has recently been published by Professor Tucker Abbott (see below for reference) who estimates that there are about 60 living species. Abbott lists 21 species and several subspecies from Australian waters and most of these are illustrated in this book.

A large proportion of cassids live in tropical seas, but there are many species in temperate seas and the family is well represented on all the shores of Australia. They are sand-burrowing molluscs, and some Australian species may be found on intertidal sand flats, but most are obtained from trawlers operating at depths to 100 fathoms or more.

Cassids are carnivorous and are thought to feed exclusively on sea urchins, biscuit urchins and sand dollars (i.e. spiny animals of the Class Echinoidea, Phylum Echinodermata). The soft parts of an echinoid's body are encased within a "test" of hard articulated calcareous plates attached to which are hard, often long and sharp spines.

Some sea urchins such as species of the genus *Diadema* have particularly long brittle and needle-sharp spines which carry a poison and are capable of inflicting painful wounds. Such creatures may seem strange prey for a soft-bodied mollusc, yet several species of the family Cassidae are known to feed on *Diadema*. Details of the feeding behaviour vary according to the species of both the cassid predator and the echinoid prey. In general the mollusc is able to immobilize the spines by squirting a paralyzing salivary juice on the urchin. It then lies on top of the urchin, apparently unaffected by the spines, and pushes its proboscis either through a hole rasped in the test by the radula or through the anus which is an unprotected part of the urchin's body. Having thus gained access to the succulent soft organs of the prey, the mollusc quickly consumes them. The entire process including the initial attack, the rasping of the hole and the ingestion of the food may take little more than an hour. Feeding seems to be done mainly at night and often while predator and prey are buried in the sand. Most published accounts of the feeding of cassids refer to Atlantic species, but it is likely that Australian members of the family also feed on echinoids. Observations on the feeding habits of Australian species would be useful.

Not a great deal is known about the breeding habits of cassids. The females lay large egg-masses (see plate 47, page 68) containing hundreds or even thousands of gelatinous capsules. In some species the egg-mass is a symmetrical tower-like structure, but in others (e.g. *Phalium glaucum, P. labiatum*) it is a large irregular mass which seems to be the work of several females spawning together. Each capsule contains several hundred eggs. Egg-laying and larval development varies among the species. In some every egg develops (e.g. *P. labiatum*), but in others only a few eggs in each capsule develop and the remainder serve as food or "nurse eggs". Early development to a shelled veliger stage takes place within the capsule, followed by a free-swimming larval stage. Sexual dimorphism in size sometimes occurs in this family; males are often smaller than females.

Cassid shells are usually oval or globose with a rather wide and long aperture. The outer lip is thickened to form a varix and its inner edge is often toothed. There may be other varices present representing the outer lip of earlier growth stages. On the left side of the aperture there is a flat or inclined shield-like callus which may be confined to the columellar area, when it is called the columellar shield, or it may cover the parietal wall as well and form a flange along the whole left side.

Behind the siphon canal, a spiral channel runs under the columellar shield into the true umbilicus. There is often another smaller cavity, the false umbilicus, under the columellar shield at the extreme anterior end. The shells of cassids may be glossy-smooth or heavily sculptured with ribs and nodules. The horny semicircular operculum does not fill the aperture.

Some species are extremely variable in shape, colour-pattern and sculpture and it is often difficult to decide on the limits between one species and another. These difficult cases have been dealt with in quite different ways by different authors. Perhaps the most notable example of this is the *Phalium pyrum* species-complex of southern Australia, New Zealand and South Africa. Older authors recognized about a dozen "species" in this complex on the basis of differences in shell characters, but in his recent monograph Abbott has shown that each of these "species" is only an extreme form, and that specimens with intermediate characters are common. He suggests that at least some of these forms and intermediates may be hybrids between properly distinct species. However, because of our lack of biological knowledge of these animals, Abbott has adopted the procedure of calling them all one species, but has retained the names originally given to the different forms as form names without formal taxonomic status. They cannot even be regarded as subspecies. This controversial procedure is an honest statement of our ignorance of the genetic relationships between the forms and, in the opinion of these authors, it should be adopted by Australian collectors unless or until biological studies show a way in which it can be corrected. The generic and subgeneric classification used here follows the arrangement given by Abbott.

Reference:

ABBOTT, R. TUCKER (1968). The helmet shells of the world (Cassidae). Part I. Indo-Pacific Mollusca, 2 (9) : 15-201, 186 pls.

1, 1a. *Cassis (Cassis) cornuta* LINNAEUS, 1758
Very solid, heavy, rotund. Spire very short, nucleus usually corroded. Surface of whorls pitted. Body whorl with 3 or 4 thick nodulose spiral ribs. Aperture narrow, outer lip broad and flat with an up-turned shelf-like rim, and 5 to 7 large teeth on its inner edge. Parietal-columellar shield broad and thin, extends around left margin to join the outer lip at the posterior end, so forming a wide flat base. Columella bears strong, irregular plicae. True umbilicus usually sealed, false umbilicus open and deep. Dorsal side and spire grey or light brown, upper edge of outer lip with 6 or 7 brown patches, base, including outer lip and parietal-columellar shield glossy-cream or orange, teeth white.
  35 cm. Common. Indo-West Pacific; Onslow, W.A. to eastern Qld. This is the largest living cassid and one of the largest gastropods. The colour, shape of the parietal-columellar shield and strength of the dorsal knobs are variable. Male shells are smaller than females and have fewer and larger dorsal knobs.

2, 2a. *Cypraecassis (Cypraecassis) rufa* LINNAEUS, 1758
Very solid, heavy, ovate. Spire short, 5 nuclear whorls. Post-nuclear whorls of spire sculptured with fine nodulose ribs. Body whorl with rounded shoulders, bears 3 spiral rows of rounded knobs (posteriorly), 1 spiral row of axially-elongated knobs (centrally), 2 spiral rows of raised axial bars (anteriorly), and between these 2 or 3 nodulose spiral ribs. Aperture narrow; outer lip very thick, with heavy teeth. Columellar and parietal areas covered by a thick glazed callus which forms a thick rim along left side of the base. On the left side strong lirae present on the inner part of the base. Inner margin of columella with an axial swelling bearing heavy teeth. True umbilicus small; false umbilicus slit-like or nearly sealed. Dorsal side orange-brown or reddish with grey to white patches, base creamy-orange, parietal wall, columella and interior dark orange-red.
  18 cm. Common in some parts of the species range but uncommon in Australia. Indo-West Pacific; northern Qld. This is the shell from which cameos are cut.

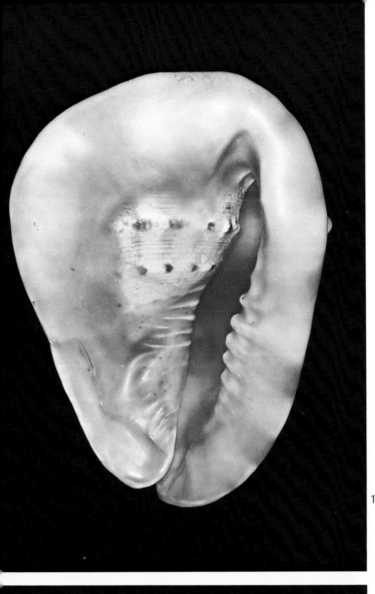
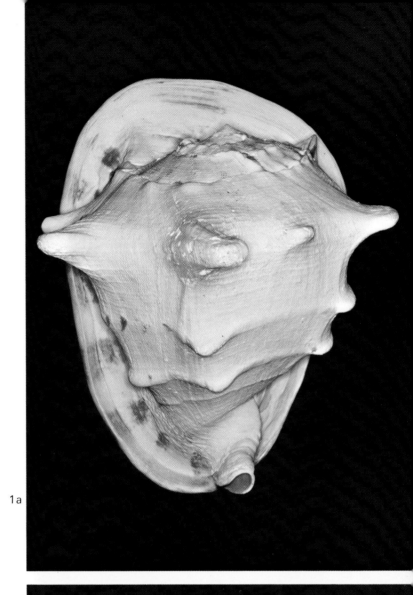

1   1a

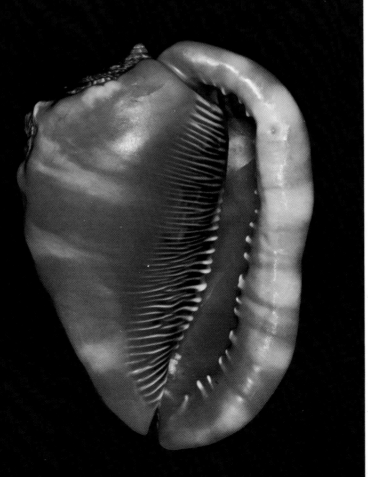
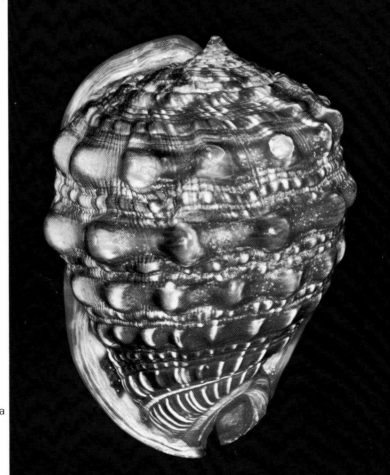

2   2a

Cassidae (Cont.)

1, 1a, 1b. *Cassis (Hypocassis) fimbriata*
QUOY & GAIMARD, 1833

Ovate to globose, spire moderately short, usually with a rather bulbous nuclear whorl at the apex. Post-nuclear whorls of spire sculptured with wavy axial ribs and fine spiral striae. Body whorl sculptured with strong but irregular axial ribs, a number of spiral cords at the anterior end, and usually 2 or 3 spiral rows of nodules or tubercles on and in front of the shoulder. Columellar shield broad, sometimes confined to columellar region but often covering parietal wall as well, smooth except for 4 or 7 strong, short marginal lirae and 4 or 5 short anterior folds. True umbilicus deep, false umbilicus small, often sealed. Outer lip usually smooth but sometimes toothed, widest near the centre. Surface shiny, cream, sometimes with axially-elongated brown patches and interrupted brown spiral lines, interior brown.

12 cm. Moderately common. Western Vic. to Houtman Abrolhos, W.A. Specimens from the northern end of the species range in W.A. are usually pink (1b). *Cassis bicarinata* JONAS, 1839 is a synonym widely but incorrectly used for this species.

2, 2a. *Cassis (Hypocassis) nana* TENISON-WOODS, 1879

Small, moderately thin, wide at the shoulders, narrow anteriorly. Spire smooth, with a small but bulbous nucleus of 1-½ whorls. Body whorl smooth, sculptured only by fine growth striae, 2 spiral rows of short sharp nodules on the shoulder and usually 1, 2 or 3 spiral rows of low weaker nodules nearer the anterior end. Columellar shield thin, sharp-edged, indented near the centre giving a gnarled appearance, columellar shield and parietal wall strongly toothed. Umbilicus sealed by columellar shield. Exterior shiny, light yellow-brown with faint brown spots between whitish nodules on the dorsal side, and faint brown radial rays on the spire, aperture teeth white, deep interior pale yellow-brown.

6 cm. Moderately common. Southern Qld. and northern N.S.W.

3, 3a. *Phalium (Phalium) glaucum* LINNAEUS, 1758

Ovate to globular. Spire high, pointed, nucleus of 2½ whorls. Early post-nuclear whorls rounded and sculptured with fine beaded spiral and axial ribs; later spire whorls with more angular nodulose shoulders and weaker sculpture. Body whorl smooth except for weak shoulder nodules and rows of shallow dents. Outer lip thickened, with sharp teeth along its inner edge and 3 or 4 sharp projecting spines anteriorly on its outer edge. Columellar shield broad, flared, and crossed by many strong but irregular lirae. True umbilicus deep, false umbilicus small, slit-like. Ash-grey, varices whitish with tan or pale orange blotches; anterior canal white with a brown tip; interior brown.

12 cm. Not common in Australia but common elsewhere within the species range. Indo-West Pacific; northern Qld.

4, 4a. *Phalium (Phalium) bandatum* PERRY, 1811

Elongate-ovate. Spire high, pointed, nucleus of 2 whorls. Early post-nuclear whorls rounded and sculptured with fine beaded axial and spiral ribs; later whorls more angulate with nodulose shoulders. Body whorl smooth except for shoulder nodules. Outer lip thickened, with sharp teeth along its inner edge and 2 or 3 poorly developed spines on the outer edge anteriorly. Columellar shield moderately broad and flaring, crossed by many strong wavy folds. True umbilicus deep; false umbilicus small. Cream to white with 5 spiral bands of rather indistinct squarish yellow-brown blotches in front of the shoulder, and a band of yellow-brown flames behind. Lower surface of lip, with 6 tan or orange blotches, anterior canal dark brown, interior light brown.

Plate 47. *Phalium (Xenophalium) labiatum* PERRY, 1811. With egg capsules at a depth of 40 feet, N.S.W., (see page 70).
(*Photo*—Neville Coleman).

12 cm. Common. Central Indo-West Pacific, Houtman Abrolhos, W.A. to northern N.S.W. Resembles *P. glaucum* but the yellow-brown blotches, higher spire and more elongate shape distinguish it from that species. *P. bandatum* is much more common and widespread in Australia than *P. glaucum.*

5, 5a. *Phalium (Phalium) areola* LINNAEUS, 1758
(see also colour plate of live animal, page 7, plate 2).

Ovate. Spire high with concave sides, a fine pointed apex, and a small nucleus of 2 whorls. Post-nuclear whorls with rounded shoulders sculptured with axial and spiral ribs and varices. Body whorl glossy, smooth except for a few weak spiral cords anteriorly. Shoulders smooth and rounded. Outer lip thickened, with strong sharp teeth along its inner edge, outer edge smooth. Columellar shield thick, moderately wide, crossed by strong lirae. True umbilicus narrow, false umbilicus small or sealed. White with 5 spiral bands of large squarish brown spots on the body whorl and along the upper edge of the lip and varices, anterior canal white, interior white although external spots may be indistinctly visible internally.

9 cm. Common. Indo-West Pacific; Broome, W.A. to northern N.S.W. Lack of shoulder nodules and the much more distinct spots readily distinguish this species from *P. bandatum.*

6. *Phalium (Semicassis) adcocki* SOWERBY, 1896

Ovate. Spire of medium height, nuclear whorls broad and smooth. Post-nuclear whorls with spiral ribs crossed by axial ribs, short axial folds also present on posterior part of body whorl. Outer lip smooth, thick. Columellar shield narrow but thick. True umbilicus almost sealed by columellar shield; false umbilicus absent. White with 6 spiral rows of small red-brown spots on the body whorl.

4.5 cm. Few specimens known. Vic. to south-eastern coast of W.A.

7. *Phalium (Semicassis) semigranosum* LAMARCK, 1822

Elongate-ovate. Spire moderately high with 2½ nuclear whorls. Post-nuclear whorls and shoulder of the body whorl with strong beaded spiral cords, remainder of the body whorl smooth or with fine spiral striae. Outer lip smooth, narrow. Columellar shield very small but thick, smooth except for weak lirae on the inner margin. True umbilicus almost sealed by columellar shield, false umbilicus absent. Exterior uniformly cream or pinkish, interior, lip and columella white.

6 cm. Common. Vic. and Tas. to about Fremantle, W.A.

8. *Phalium (Semicassis) sophia* BRAZIER, 1872

Differing from *P. bisulcatum* mainly in the tabulate shoulders, more-or-less smooth surface, and broader columellar shield.

8 cm. Moderately common. Southern Qld. and N.S.W. Abbott, (1968) places this form as a subspecies of *P. bisulcatum* though earlier authors generally accepted it as a distinct species. It may be a hybrid form between *P. pyrum* and *P. bisulcatum.*

9, 9a. *Phalium (Semicassis) bisulcatum*
SCHUBERT & WAGNER, 1829

Thin or thick-shelled, ovate to globose. Spire of medium height, with very small nucleus of about 3 whorls. Early post-nuclear whorls rounded and sculptured with beaded spiral and axial ribs, axial ribs become obsolete on later spire whorls. Body whorl smooth or with strong flattened spiral ribs or fine striae. Outer lip moderately thick, with teeth along its inner edge but the outer edge smooth. Columellar shield narrow, crossed by strong lirae. False umbilicus and true umbilicus deep. A deep spiral channel lies behind the anterior canal. White, cream, blue-grey or pink, often with 5 or 6 spiral rows of reddish or yellow-brown spots which are usually rather square. Upper edge of lip varix spotted with brown, inner edge of lip and columellar shield white, interior white.

7 cm. Moderately common. Indo-West Pacific; Shark Bay, W.A. to southern Qld.

10, 10a. *Phalium (Semicassis) glabratum angasi* IREDALE, 1927

Thin, globose. Spire of medium height, nucleus of 3 whorls. Early post-nuclear whorls with beaded spiral and axial threads, later spire whorls smooth. Body whorl with rounded and faint squarish dents. Outer lip narrow, with widely-spaced fine sharp teeth along its inner edge. Columellar shield narrow, margin indented, strongly lirate. False umbilicus and true umbilicus of almost equal size, deep and round; a pointed projection of the columella borders the anterior inner edge of the false umbilicus. Translucent white, pale tan or pale pink, sometimes with a few small brown blotches near the sutures.

6 cm. Moderately common. Kimberley region, W.A. to northern N.S.W. This subspecies is confined to Australia; the typical form *P. glabratum glabratum* DUNKER, 1852 is a rare mollusc from the Indo-Malay Archipelago.

Plate 48. Cassidae (⁸⁄₁₀ × natural size)

Cassidae (Cont.)

1, 1a, 1b, 1c. *Phalium (Xenophalium) pyrum* LAMARCK, 1822

This is a problematical species. Many forms of it have been recognized as distinct species, but Abbott (1968) provisionally considers most of them to be ecological forms having no taxonomic status. Typical specimens of the better known Australian forms are illustrated here and their original names are used as forma following Abbott. Other and similar forms occur in South Africa and New Zealand.

The typical form named *pyrum* by Lamarck (not illustrated): Globose to ovate. Spire of medium height, with 3 nuclear whorls. Post-nuclear whorls with strong spiral striae and very fine axial striae, later spire whorls with a finely nodulose spiral rib on the shoulders. Shoulder nodules weak or absent on the body whorl, but in front of the suture are a distinct incised spiral groove and sometimes weak spiral striae, also 3 or 4 spiral grooves at the anterior end. Outer lip narrow, its inner edge smooth. Columellar shield small, thick, smooth except for a few weak marginal lirae. True umbilicus deep, false umbilicus very small or sealed. Fawn with spiral bands of diffuse light brown spots or stains, anterior canal tipped with a brown patch, columella and inner edge of outer lip white, interior brown.

7 cm. Moderately common. Recorded from N.S.W., Vic., Tas. and S.A. Also found in New Zealand and South Africa.

1b, 1c. *forma stadiale* HEDLEY, 1914. Thinner, more elongate and higher-spired than typical *P. pyrum*. Shoulders smooth, pre-sutural incised groove and anterior spiral grooves very weak. Inner columellar margin bears several moderately strong oblique lirae anteriorly.

9 cm. N.S.W. to the south-eastern coast of W.A. Specimens described by Cotton (1945, 1954) as *Xenogalea denda* and *X. halli* are of this form.

1. *forma niveum* BRAZIER, 1872. Similar to typical *P. pyrum* but the shoulders more angulate and bear 2 spiral rows of nodules or tubercles. Colour uniform cream or fawn.

9 cm. N.S.W. to Bremer Bay in W.A.

1a. *forma spectabile* IREDALE, 1929. Intermediate between typical *P. pyrum* and the *niveum* form.

9 cm. Recorded from depths to 120 fathoms, N.S.W. to the south-eastern coast of W.A. Abbott (1968) considers that *Xenogalea mawsoni* COTTON (1945) from the south-eastern coast of W.A. may be assigned to this form.

2, 2a. *Phalium (Xenophalium) pauciruge* MENKE, 1843

Ovate. Spire high, pointed, with a small nucleus of 2½ whorls. Early post-nuclear whorls rounded and sculptured with fine spiral striae and axial ribs, later spire whorls usually with weakly nodulose shoulders. Body whorl smooth except for shoulder nodules which become obsolete toward the lip. Outer lip thick, usually with teeth along its inner edge, outer edge smooth. Columellar shield thick and smooth except for short folds along the inner margin. True umbilicus moderately deep but partly closed, false umbilicus rarely present. White or yellowish with yellow-brown flames in the sutures and 4 bands of small widely-spaced indistinct yellow-brown spots. Outer lip and columellar shield white, anterior canal white, tipped with brown, interior yellow-brown.

7 cm. Common. Southern W.A. from Esperance to Shark Bay.

3, 3a, 3b. *Phalium (Xenophalium) thomsoni* BRAZIER, 1875

Globose. Spire high, pointed, with 4 nuclear whorls. Post-nuclear whorls shouldered and sculptured with fine axial and spiral striae. Body whorl usually shouldered, sculptured with 1 to 3 spiral rows of rounded or axially elongated nodules, fine spiral striae and a few stronger incised lines anteriorly. Outer lip strong, toothed on inner edge. Columellar shield narrow, strongly and irregularly plicate, outer margin strongly indented. True umbilicus deep, false umbilicus small or closed. White, cream or tan, usually with 3 to 5 spiral rows of red-brown spots, outer lip and columella white.

9 cm. Moderately common. Southern Qld. to Vic. and Tas. The number and strength of the shoulder nodules are variable. Iredale (1931) gave the name *palinodium* to an extremely nodulose dwarf form.

4, 4a. *Phalium (Xenophalium) whitworthi* ABBOTT, 1968

Ovate. Spire very high and pointed, with 4 nuclear whorls, early post-nuclear whorls sculptured with spiral ribs and axial threads, later spire whorls sculptured with strong spiral ribs. Body whorl with 12 to 14 very wide and strong nodulose spiral ribs. Outer lip thick, bearing 15 to 20 teeth (often paired) along its inner edge, outer edge smooth. Columellar shield short, narrow but strong, strongly rugose. False umbilicus and true umbilicus deep. Cream or pale tan with a few small light brown spots, outer edge of lip varix and tip of anterior canal spotted with darker brown, columellar shield and inner edge of lip white, interior yellow-brown.

8 cm. Uncommon. W.A. between Rottnest Is. and Geraldton. Dead shells with hermit crabs are often taken in craypots, but living specimens are only rarely dredged.

5. *Phalium (Xenophalium) labiatum* PERRY, 1811

(see also colour plate of live animal, page 68, plate 47)

Globose to oblong-ovate. Spire of medium height. Whorls smooth and rounded or with nodulose shoulders. Outer lip narrow, usually toothed along the inner margin. Columellar shield small, thick and smooth except for weak lirae on its inner margin. True umbilicus deep but narrow, almost sealed, false umbilicus very small or sealed. Exterior shiny, bluish and cream with purple-brown splashes sometimes V-shaped and joined axially to form zigzag axial lines, especially on the parietal wall. Spiral row of brown spots present in front of the sutures and dark brown blotches on the outer lip, columella orange-yellow, anterior canal tipped with brown, interior light brown.

8 cm. Moderately common. Southern Qld. to western Vic. Also found in northern New Zealand and Norfolk Island. *Xenogalea imperata* IREDALE (1927) from N.S.W. is a form of *P. labiatum* with nodulose shoulders.

6. *Casmaria ponderosa* GMELIN, 1791

Oblong-ovate. Spire of medium height, spire whorls smooth, rounded, body whorl smooth, shoulders rounded and either smooth or nodulose. Outer lip thick with 1 or 2 rows of widely spaced tooth-like denticles along its entire length. Columellar shield small, thick, with many fine irregular lirae. True umbilicus sealed, false umbilicus absent. White, cream or tan with a spiral row of brown spots below the sutures and another behind the anterior canal. Outer lip bears large dark brown spots, a dark brown blotch lies in base of siphonal notch, lip and columella white, interior brown.

5 cm. Common. Indo-West Pacific; Barrow Is., W.A. to southern Qld. Two forms of this species occur throughout its range, one thin, smooth and without shoulder nodules (6), and the other heavy with shoulder nodules (not illustrated). *C. ponderosa perryi* IREDALE, 1912 is a cool water subspecies which differs from typical *C. ponderosa* in having 3 or 4 indistinct light brown spiral bands in addition to the pre-sutural and anterior bands of blotches, and in lacking denticles along the anterior third of the outer lip. It is found in N.S.W., northern New Zealand, Kermadec Islands and Easter Is., South Pacific.

7, 7a. *Casmaria erinacea* LINNAEUS, 1758.

Oblong-ovate. Spire of medium height; spire whorls rounded. Body whorl smooth, shoulders rounded and either smooth or nodulose. Outer lip thick, smooth along its inner edge bears several prickle-like teeth on the outer edge at its anterior end. Parietal wall glazed with callus, columellar shield thick, small, with a spiral swelling centrally, 1 strong spiral fold anteriorly and several weak marginal spinal lirae behind that. True umbilicus sealed, false umbilicus absent. Whorls white, grey, cream or pale tan, sometimes with irregular bands of diffuse brown blotches, usually with axial rows of minute brown spots. Outer lip white with large dark brown spots. A dark brown blotch lies in the base of the siphonal notch. Columella white, interior brown.

7 cm. Common. Central Eastern Pacific and Indo-West Pacific; Pt. Cloates, W.A. to eastern Qld. Two forms of *C. erinacea* occur throughout its range. One of these is rather thin-shelled and has smooth shoulders (7), the other is heavy and has nodulose shoulders (7c). *C. ponderosa* has 2 similar forms, and the 2 species are sometimes hard to tell apart. *C. ponderosa* lacks the axial rows of minute spots typical of *C. erinacea* but has a spiral pre-sutural row of brown spots and another spiral row of brown spots behind the anterior canal. The anterior teeth on the outer edge of the lip are also usually lacking in *C. ponderosa* but there are usually fine denticles along the inner edge of the lip throughout its length.

Plate 49. Cassidae (des. page 70) and Ficidae (des. page 72) (⁹⁄₁₀ × natural size)

fig shells (FAMILY FICIDAE)

FIG SHELLS ARE THIN AND DELICATELY SCULPTURED, and take their vernacular name from their shape. The spire is low, the body whorl inflated and the anterior canal is long, tapered and gracefully curved. There is no operculum and the aperture is wide. The animal has a small head, a long siphon and a large foot, and the mantle lobes spread over most of the shell. Fig shells are active creatures which live in sand and feed on sea urchins or other echinoderms. There are less than a dozen species in the family, all of them living in tropical or warm temperate seas. Four species are recorded from the western, northern and eastern coasts of Australia.

Refer to page 71, plate 49, for (8, 8a, 9, 9a) colour illustrations of shells.

8, 8a. *Ficus tesselata* KOBELT, 1881.

Spire nearly flat, sutures slightly indented, apex protruding and consists of a single smooth bulbous nuclear whorl. Whorls cancellate, with strong primary spiral ribs and slightly weaker intermediate and axial ribs. Usually only one intermediate rib in each interspace. Cream or white with 6 to 10 spiral rows of squarish brown spots.

    5 cm. Common. North-West Cape to Broome, W.A. Literature records of this species from the Philippines and other Indo-West Pacific localities need confirmation.

9, 9a. *Ficus subintermedia* D'ORBIGNY, 1852

(see also colour plate of live animal, page 2, *frontispiece*)

Spire slightly elevated, sutures indented, nucleus of 2 smooth whorls. Whorls cancellate, with strong primary spiral ribs, weaker secondary spiral ribs, and often 1 or more weak intermediate spiral ribs in the interspaces. Axial ribs about as thick as the secondary spiral ribs. Pale brown with narrow paler brown or cream spiral bands, and irregular spiral rows of diffuse brown spots.

    8 cm. Common. Indo-West Pacific; Onslow, W.A. to northern N.S.W. In Australian literature this species has been called *F. ficoides* LAMARCK, 1822 (non BROCCHI, 1814) and *F. communis* RÖDING, 1798, but neither name is correct. *F. ficus* LINNAEUS, 1758 is a much broader shell with less cancellate sculpture widespread in the Indo-West Pacific and recorded from Barrow Is., W.A. *F. filosa* SOWERBY, 1892 is a thin strongly ribbed Indo-West Pacific species recorded from N.S.W.

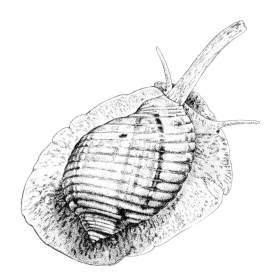

Fig. 15. *Tonna chinensis* DILLWYN, 1817, drawn from colour photograph of a living animal, Margaret River, W.A.

tun shells

(FAMILY TONNIDAE)

THE SHELLS OF THIS FAMILY are generally large, rather thin and inflated, with strong spiral ribs but no axial sculpture or true varices, and bear a thin smooth greenish periostracum. The anterior canal is a U-shaped notch. There is often a columellar callus which may be reflected over the umbilicus and parietal wall to form a shield. In most forms the umbilicus is deep but is lacking in the genus *Malea*. This genus also differs from other tuns in having a solid shell with a thick, heavily toothed outer lip, and with strong folds and a deep notch on the columella.

Tuns are sand dwellers. Most species live in deeper water and are rarely found alive in the intertidal zone. The body of the animal is very large when fully extended. It has a wide, rather flat foot, a long respiratory siphon, prominent eye stalks and a long eversible proboscis. Although the body can be withdrawn within the protection of the shell there is no operculum to shut it in. When placed in aquaria living specimens quickly bury themselves with only the tip of the respiratory siphon showing above the surface of the sand. Presumably this is the resting position, and it seems likely that the animals crawl on the surface of the sand only when hunting for food. Very little is known about the feeding habits of tuns, they are certainly carnivores but the nature of their prey is unknown. Neither is there much information on their breeding habits although they are known to have very long-lived pelagic larval stages.

The family is confined to tropical and warm temperate seas and is most strongly represented in the Indo-West Pacific. The exact number of Australian species is uncertain but most of them are illustrated or referred to here. Taxonomically the tuns are a very confusing group. There has been no revision of the family for many years. Hedley (1919, Rec. Aust. Mus. **12** (11) : 329-336, pls 39-44) published the last account of the Australian species but this is now outdated.

Plate 50. *Tonna perdix* LINNAEUS, 1758, Qld.   (*Photo*—D. Henderson).

1, 1a. *Tonna variegata* LAMARCK, 1822

Spire with 3 smooth nuclear whorls. Post-nuclear whorls (about 5) all sculptured with broad rounded spiral ribs (about 17 on the body whorl) and weaker interstitial riblets on the posterior parts of the whorls. Columella reflected over wide spiral umbilicus, anterior fasciole broad. Yellow, cream or light brown, with at least some spiral ribs white or pale cream, and sometimes with broad brown spots.

20 cm. Moderately common at depths down to at least 100 fathoms. Eucla to Shark Bay, W.A.

Tonna cerevisina HEDLEY, 1919 (not illustrated), which is trawled off the east coast of Australia and New Zealand is very like *T. variegata* but lacks the interstitial riblets at the posterior ends of the whorls. Hedley (1919) illustrated a W.A. specimen of *T. variegata* with 4½ whorls which was only 9.5 cm in length, while an eastern specimen with this number of whorls, which he named *T. cerevisina* was 17 cm in length. However, this difference between eastern and western specimens is not consistent. Study of series of specimens from intermediate southern Australia localities may show *T. variegata* and *T. cerevisina* to be extreme end-of-range forms of a single species.

Tonna tetracotula HEDLEY, 1919 is another large and similar tun from eastern Australia and New Zealand. It is often uniform white to pale orange, but sometimes has light brown spiral bands. Otherwise it may be distinguished from *T. cerevisina* by its higher spire and the presence of riblets between the ribs on the shoulder.

Fig. 16. *Tonna perdix* LINNAEUS, 1758. Sectional view.
(*Photo*—Keith Gillett).

2. *Tonna allium* DILLWYN, 1817

Resembles *T. tessellata* but spire slightly higher, ribs usually fewer, and exterior cream or fawn with slightly darker ribs lacking spots.

10 cm. Moderately common. This species has been found on intertidal sand flats in Exmouth Gulf but also lives in deeper water. Indo-West Pacific; Exmouth Gulf, W.A. to northern N.S.W. Another similar species is *T. ampullacea* PHILIPPI, 1845 which has strong riblets between the spiral ribs. Some authors have regarded *T. ampullacea* as a form of *T. allium* but Hedley (1919) believed it to be a distinct species. The widely used name *T. costata* MENKE, 1828 is a synonym of *T. allium*.

3. *Tonna chinensis* DILLWYN, 1817

(see Fig. 15, page 72).

Spire with a nucleus of 3 smooth whorls and shallow sutures. Post-nuclear whorls 4, sculptured with strong rounded spiral ribs (about 17 on body whorl) separated by grooves slightly narrower than the ribs. Anterior fasciole wide and raised. Columella thin, reflected over a wide umbilicus, anterior canal notch deep, oblique. Light yellow-brown or cream with white dashes and brown spots on the ribs, interior yellow-brown becoming white near the lip.

6 cm. Common in shallow water and sandy pockets of intertidal reefs. Indo-West Pacific; Cape Leeuwin, W.A. to N.S.W. Specimens of this species are often misidentified as immature *T. variegata*, but they lack interstitial ribs and are consistently smaller for any given number of whorls.

4, 4a. *Tonna perdix* LINNAEUS, 1758

(see also colour plate of live animal, page 73, plate 50).

Spire high, with incised sutures and a nucleus of 2 smooth whorls. Post-nuclear whorls (about 5) sculptured with about 18 broad low rounded spiral ribs. Anterior canal notch wide, very shallow and oblique, anterior fasciole broad but low. Columellar margin gently curved, columellar callus thin and reflected over a deep umbilicus. Light brown with white lines in the grooves between the ribs and numerous crescent-shaped white markings on the ribs, deep interior yellow-brown becoming white near the lip, columellar callus white.

14 cm. Common. Indo-West Pacific; Houtman Abrolhos, W.A. to eastern Qld.

5, 5a. *Tonna tessellata* LAMARCK, 1816

Spire with a nucleus of 2 smooth whorls. Post-nuclear whorls with broad rounded spiral ribs (about 14 on the body whorl), inter-spaces wide and smooth. Outer lip slightly reflected, crenulate. Outer wall of aperture deeply spirally grooved by impression of external ribs. Columella calloused, callus reflected and usually extended posteriorly over the parietal wall. Umbilicus small but deep. Fawn or blue-grey between the ribs, ribs white or pale fawn with large squarish brown blotches, interior brown.

10 cm. Moderately common. Indo-West Pacific; Dampier Archipelago, W.A. to eastern Qld.

6. *Tonna canaliculata* LINNAEUS, 1758

Spire with 3 smooth nuclear whorls, and widely and deeply channelled sutures. Post-nuclear whorls (about 5) sculptured with wide low spiral ribs (about 17 on body whorl) which are separated by narrow grooves, posterior ribs on the body whorl flat, anterior ribs rounded. Columella vertical, almost straight, reflected over a deep umbilicus. Anterior fasciole broad but low. Anterior canal notch wide, shallow, oblique. Light yellow-brown, darker on the ribs than in the grooves, with brown and white blotches at the sutures. Deep interior yellow becoming white near the margin.

12 cm. Uncommon. Indo-West Pacific; Barrow Is. W.A. to eastern Qld.

7. *Malea pomum* LINNAEUS, 1758

Spire with a nucleus of 3 smooth whorls and weakly incised sutures. Post-nuclear whorls 4, sculptured with strong, rounded, spiral ribs (about 12 on body whorl), separated by wide, shallow grooves. Outer lip thickened, with about 10 strong teeth on its inner margin. Teeth at the anterior end continuing across the lip to form outward pointing spines. Columella deeply notched and calloused, with strong marginal lirae. Anterior canal notch deep, oblique. Fawn or cream-orange with large white spots on the ribs, interior yellow, columella and outer lip white or cream.

6 cm. Common. Indo-West Pacific; Houtman Abrolhos, W.A. to eastern Qld.

Plate 51. Tonnidae (⁸⁄₁₀× natural size)

tritons and trumpet shells (FAMILY CYMATIIDAE)

MANY COLOURFUL SPECIES OF THIS FAMILY live in Australia's tropical north and a few in the temperate waters of the south. It has been possible to illustrate only a few examples in this book.

Cymatiid shells usually have strong varices, teeth on the outer lip and folds or ridges on the columellar side of the aperture. In most species the shell is thick, heavy and strongly sculptured with nodules and spiral ribs, and there is a hairy, scaly or bristly periostracum and a horny operculum. Often the thick periostracum must be removed from the shell before the details of colour and sculpture can be seen. Although these general shell characters apply to most cymatiids, shell form is very varied in this family and some of the most important diagnostic characters of the family are anatomical.

Charonia tritonis of the tropical Indo-West Pacific is notable for several reasons. Apart from its outstanding beauty it is the second largest of the world's living gastropods, individuals growing to a length of nearly two feet.

That *Charonia tritonis* preys upon the Crown of Thorns and other starfish has been discussed in the Introduction. *Charonia rubicunda* also feeds on starfish. Many other cymatiids (e.g. *Mayena australasia, Monoplex australasiae, Cabestana spengleri*) eat mainly simple ascidians (John Laxton personal communication). Other species such as *"Cymatium" nicobaricum, "C." gemmatum, "C." pileare* and *"C." muricinum* eat other molluscs which they paralyze with a strongly acid saliva (Houbrick and Fretter, 1969). Cymatiids lay their egg capsules in cup-shaped clusters in crevices or on the underside of stones. The larval development of a few species has been studied, and in most of these the embryos hatch at a late free-swimming veliger stage. Scheltema (1966) reported that the Caribbean species *Cymatium parthenopeum* VON SALIS, 1793 has a planktonic larval life of about three months. If such a long planktonic larval stage is general in the family it may account for the exceptionally wide distribution of some species. However at least one species, *Cabestana spengleri,* has been reported to have direct development (Anderson, 1959).

A number of cymatiids (e.g. *Lampusia nicobaricum, L. pileare* and *Charonia tritonis*) occur in the tropical western Atlantic as well as in the Indo-West Pacific region. Others belong to closely related groups of forms or species which have representatives in widely separated isolated parts of the world. Such species groups present difficult taxonomic problems. For example, the Australian species *Charonia rubicunda* of the south east and *C. powelli* of the south west, belong to a group which includes *C. lampis* of the Mediterranean, *C. Sauliae* of Japan and *C. capax* of New Zealand. Shells of these isolated populations are all alike and interpretation of their taxonomic relationships is difficult.

Selected references:

ANDERSON, D.T. (1959). The reproduction and early life history of the gastropod *Cymatilesta spengleri* (Perry) (Fam. Cymatiidae). Proc. Linn. Soc. New South Wales, **84** : 232-237.
DALL, W. H. (1904). An historical and systematic review of the Frog shells and Tritons. Smithsonian Institution Miscellaneous Collections, Historical Series, **47** : 114-144.
HOUBRICK, JOSEPH R. & FRETTER, VERA (1969). Some aspects of the functional anatomy and biology of *Cymatium* and *bursa*. Proc. malac. Soc. Lond., **38** : 425-429.
LAXTON, J. H. (1969). Reproduction in some New Zealand Cymatiidae (Gastropoda : Prosobranchiata). Zool. J. Linn. Soc., **48** : 237-253, pls 1-2.
SCHELTEMA, R. S. (1966). Evidence for trans-Atlantic transport of gastropod larvae belonging to the genus Cymatium. Deep Sea Res., **13** : 83-95.

1. *Charonia tritonis* LINNAEUS, 1758

Body whorl swollen, spire high and pointed, varices broad but low, anterior canal short. Sculptured with smooth, wide, rather flat spiral ribs, and a single narrow axially-plicate rib at the sutures. Outer lip bears 9 to 12 pointed teeth; columella with strong bifid lirae. Fawn or light brown with crescent-shaped markings of purple-brown, interior orange-yellow or reddish, columella dark brown in the interstices between white lirae.

45 cm. Uncommon. Indo-West Pacific; Dongara, W.A. to eastern Qld.

2. *Charonia rubicunda* PERRY, 1811

Spire high and pointed, varices weak and thin, anterior canal short. Sculptured with low spiral ribs, 2 or 3 central ribs sometimes broad and nodulose, others narrow, fine axial striae in the interspaces. Outer lip expanded, denticulate. Parietal ridge long and strong, columella smooth except for a few weak, short nodules anteriorly. Yellow-brown or reddish with darker red-brown blotches, apex pink, interior white, outer lip white, teeth dark brown, columellar callus brown.

15 cm. Common. N.S.W. to S.A. Another form in deep water in the Great Australian Bight is known as *C. euclia* HEDLEY, 1914. The relationship between *C. rubicunda, C. powelli, C. euclia* and the New Zealand species *C. capax,* is apparently very close and they may be variants of a single species.

3. *Charonia powelli* COTTON, 1956

Closely resembles *C. rubicunda* but narrower, much more solid and more heavily nodulose. Columella usually strongly lirate throughout its length.

15 cm. Moderately common. Southern W.A. from Esperance to Fremantle.

4. *Cymatium lotorium* LINNAEUS, 1758

Varices very thick; anterior canal short, bent, slightly up-turned. Sculptured with about 4 very heavy ridged nodules between the varices, and with spiral ribs which become thick and nodulose where they cross the varices. Outer lip with broad teeth which descend as thick folds into the aperture, columella with a few short irregular folds. Parietal ridge represented by a large nodule. Exterior yellow to orange-brown or red-brown, ribs white where they cross the varices, dark brown in the interspaces. Interior white, margin of outer lip brown, parietal wall with 2 dark brown patches.

15 cm. Moderately uncommon. Indo-West Pacific; eastern Qld.

5. *Ranularia pyrum* LINNAEUS, 1758

Spire low, body whorl inflated, anterior canal long, narrow and up-turned. Sculptured with about 5 axial rows of heavy nodules between the thick varices, and with spiral ribs which become thick and heavy as they cross the varices. There are about 7 thick divided teeth on the inner margin of the outer lip, and strong irregular folds on the columella. Exterior orange-brown or orange-red, sometimes with white on the varices, columella and margin of outer lip orange with white teeth, deep interior white, Periostracum brown and thin, except on the nodules where it forms long comb-like rows of bristles.

10 cm. Uncommon. Indo-West Pacific; eastern Qld.

6. *Monoplex australasiae* PERRY, 1811

Broadly fusiform, varices rather weak, anterior canal moderately short. Sculptured with thick, rounded, nodulose spiral ribs, and with fine spiral and axial striae. A double row of paired teeth present on inner margin of outer lip. Columella with irregular folds. Exterior brown, deep interior white, lip teeth and columellar folds white, interstices dark brown. Periostracum thick and scaly.

14 cm. Common. Southern Qld. to Fremantle, W.A.

7, 7a. *Cabestana spengleri* PERRY, 1811

Rather fusiform, spire moderately high, anterior siphon canal short. Sculptured with strong, broad spiral ribs crossed by numerous longitudinal ridges, and with about 6 nodules on the shoulders between the varices. Varices strong, crossed by spiral ribs. Outer wall of aperture with strong paired spiral ribs which thicken near the lip to form pairs of strong tooth-like structures. Body whorl ribs usually visible beneath the posterior end of the columella, which is otherwise smooth or weakly lirate. Exterior yellow-brown covered with a thin brown periostracum, interior and columella glossy white.

15 cm. Common. Southern Qld. to Tas. and S.A.

8, 8a. *Mayena australasia* PERRY, 1811

Fusiform, varices strong, anterior canal short. Whorls more or less smooth, sculptured with fine spiral striae and a central spiral row of low nodules, which are usually obsolete on the body whorl (8) but sometimes persist (8a). Outer lip with 7 to 10 short teeth, some of which may be bifid. Parietal ridge represented by a thick nodule, columella with only a few weak folds. Whorls brown, varices banded brown and white, interior white. Periostracum thick and mat-like.

9 cm. Common. N.S.W. to southern W.A.

Plate 52. Cymatiidae ($\frac{7}{10}$ × natural size) ⇨

Cymatiidae (Cont.)

1. *Argobuccinum vexillum* SOWERBY, 1835
Fusiformly ovate, anterior canal short, projecting. Whorls rounded, sculptured by about 8 spiral rows of weak rounded nodules which sometimes form longitudinal folds especially on the earlier whorls. Varices low and rounded. Inner side of outer lip with a nodulose ridge, columella toothed anteriorly. Parietal ridge represented by a tooth-like nodule. Exterior fawn with brown nodules or spiral bands, interior white. Periostracum thick, brown and matted.
   10 cm. Moderately common. Vic., Tas. and S.A. The number and strength of the dorsal nodules is variable.

2. *Fusitriton retiolus* HEDLEY, 1914
Rather thin, ovately fusiform, spire high, anterior siphon canal moderately long and up-turned, varices low and narrow. Whorls rounded, sculptured by intersecting axial and spiral cords, with low rounded nodules at the junctions. Parietal ridge represented by a heavy nodule. Outer lip and outer wall of aperture smooth, columella smooth, sharply cornered at the base of the anterior canal. Exterior white or cream and covered by a thin yellowish periostracum, interior white, sometimes yellow on the outer lip.
   13 cm. Uncommon. Trawled off N.S.W. and Vic.

3. *Lampusia nicobaricum* RÖDING, 1798
Very solid, spire high, and shouldered, anterior canal short and up-turned. Whorls sculptured with 4 to 7 high nodulose axial folds crossed by 7 to 9 strong spiral ribs with 2 fine but raised cords in the interspaces. Varices broad and heavily nodulose, nodules formed by thickening of the spiral ribs where they cross the varices. Outer lip with strong elongate denticles (sometimes divided) on the inner margin. Columella strongly plicate. Exterior grey, flecked or blotched with brown, aperture teeth white, interstices bright orange or yellow.
   9 cm. Moderately common on shallow reefs. Indo-West Pacific; Rottnest Is., W.A. to northern N.S.W. Also found in the West Indies.

4, 4a. *Gutturnium muricinum* RÖDING, 1798
Solid and stout, spire low, shouldered, anterior canal narrow but moderately short and up-turned, varices broad and heavy. Sculptured with 4 or 5 high nodulose axial folds, crossed by strong granulose spiral ribs and striae. 6 or 7 strong denticles (sometimes divided) present on inner margin of outer lip. Columella calloused, plicate, central plications weak, anterior plications strong. Exterior usually blue-grey or fawn-grey flecked or blotched with brown, sometimes dark brown with a central white spiral band. Outer lip and columella glazed, glossy white or yellow, interior red or purple-brown.
   6 cm. Common. Indo-West Pacific; Shark Bay, W.A. to eastern Qld. Also found in the West Indies.

5, 5a. *Turritriton tabulatus* MENKE, 1843
Ovate to fusiform, spire of medium height, heavily nodulose and shouldered, anterior canal moderately long, usually pointed to the animal's right. Varices strong and nodulose. Whorls sculptured with 4 sets of heavy nodules between the varices, crossed by strong granulose spiral ribs and fine striae. Outer lip with about 6 strong elongate denticles on its inner margin, columella calloused, smooth except for a few short folds anteriorly. Exterior greyish with brown patches and spiral bands, interior white.
   7 cm. Common. All Australian states. The length of the anterior canal and the nodulosity are very variable.

6. *Negyrina subdistorta* LAMARCK, 1822
Spire high, nodulose, shouldered. Spire whorls distorted somewhat asymmetrical, varices heavy, broad. Anterior canal moderately short, up-turned. Sculptured with beaded spiral ridges, 3 larger central ridges bear nodules which are heavy on the spire and weak on the body whorl. Outer lip smooth, columella bears a number of weak folds. Exterior cream or fawn, mottled with red-brown, interior white.
   7 cm. Common. N.S.W. to S.A.

7, 7a. *Distorsio anus* LINNAEUS, 1758
Solid, ovate, grossly distorted. Outer lip and ventral callus expanded to form a flattened base with a broad, frilled rim. Spire of medium height, anterior canal short and nearly vertical, varices weak. Body whorl sculptured with heavy nodules, strong spiral ribs and fine axial striae. Heavy teeth present on the outer lip. Posterior and anterior ends of the columella project into and partially occlude the aperture. Dorsum cream or greyish with brown patches and spiral bands, ventral surface polished, mottled white and fawn, rim patched alternately brown and white.
   10 cm. Moderately uncommon. Western Pacific; eastern Qld.

8, 8a. *Distorsio reticulata* RÖDING, 1798
Rather light, fusiform, distorted. Spire moderately high and pointed, anterior canal moderately short and up-turned, varices weak. Sculptured with fine spiral and axial ribs with a small nodule at each intersection. Outer lip flat and expanded, crossed by rib-like teeth. Parietal ridge consisting of 2 prominent nodules, columella bears strong marginal teeth anteriorly. Ventral side of body whorl glazed. Dorsal surface cream or yellow with yellow-brown spiral bands, ventral side light brown, teeth white. Periostracum brown with a single long bristle from each nodule on dorsal surface.
   9 cm. Moderately common. Indo-West Pacific; Shark Bay, W.A. to eastern Qld.

9, 9a. *Austrosassia parkinsoniana* PERRY, 1811
Broadly fusiform, spire high, anterior siphon canal short and slightly up-turned, varices strong. Sculptured with crowded, very fine granulose spiral cords and 3 to 5 large knobs between the varices. Outer wall of aperture smooth except for about 6 denticles near the lip. Exterior yellow or orange-brown with spiral cords of alternating white and brown spots, interior glossy white. Periostracum thin, yellowish.
   4.5 cm. Common. N.S.W., Vic. and Tas.

10, 10a. *Septa hepatica* RÖDING, 1798
Very like *S. rubecula* but usually red-brown with 6 to 8 blackish spiral bands in the interspaces between the spiral ribs, and red-orange interstices between the outer lip teeth.
   5 cm. Uncommon. Pacific; eastern Qld.

11, 11a. *Septa rubecula* LINNAEUS, 1758
Broadly fusiform, spire of medium height, anterior canal short, varices broad and strong. Whorls sculptured with strong nodulose beaded spiral ribs, and with axial ridges in the interspaces. There are 8 to 10 strong denticles on the inner margin of the outer lip, and strong folds on the entire length of the columella. External colour variable, usually bright red, sometimes brown, with narrow yellow or white spiral bands and patches of yellow or white on the varices. Interior, outer lip denticles and columellar folds white, columella reddish between the folds, interstices between the outer lip denticles white.
   5 cm. Moderately common. Indo-West Pacific; Dampier Archipelago, W.A. to eastern Qld.

12, 12a. *Lampusia pileare* LINNAEUS, 1758
Fusiform, varices strong, anterior canal short. Sculptured with strong spiral ribs which thicken as they cross the varices, ribs nodulose tending to form axial ridges. Outer lip with 14 to 18 strong paired teeth which descend deep into the aperture as spiral lirae, columella with strong irregular folds, some of them bifid. Exterior cream, light brown or light grey, with brown spiral bands and axial streaks, interior and columella dark orange or red, teeth and columellar folds white. Periostracum thin with comb-like rows of bristles on the axial ridges and varices.
   10 cm. Common. Indo-West Pacific; Rottnest Is., W.A. to eastern Qld.

13, 13a. *Cabestana waterhousei* ADAMS & ANGAS, 1864
Spire high and shouldered, varices strong, anterior canal moderately short. Sculptured with 7 to 10 strong, raised, double spiral ribs and more numerous fine spiral lirae in the interspaces, all crossed by fine axial ribs or cords. Shoulder ribs nodulose, especially on the spire whorls. Outer lip expanded, flattened and crossed by spiral ribs and striae. Columella smooth. Exterior uniformly yellow-brown, interior white.
   10 cm. Common. N.S.W. to Fremantle, W.A.

14. *Gyrineum (Biplex) aculeatum* SCHEPMAN, 1909
Resembles *G. (Biplex) pulchellum* but larger with strong spines projecting from the expanded varices.
   4.5 cm. Uncommon in deep water. Central Indo-West Pacific; Carnarvon to Onslow, W.A. This species is even more closely related to *G. (Biplex) percum* PERRY, 1811 of Central Indo-West Pacific. Its presence in Australian waters has only recently been established.

15, 15a. *Gyrineum (Biplex) pulchellum* FORBES, 1852
Flattened, spire high, with deeply channelled sutures, varices expanded to form a broad, thin, fin-like rim along both sides. Aperture ovate to circular with a raised rim, anterior canal moderately long, narrow and tubular. Whorls sculptured by strong axial and spiral ribs with nodules at the intersections, ribs run out on to both sides of the varices. Exterior grey, cream or white, interior white.
   2.5 cm. Moderately common in sand or mud in deeper water. Central Indo-West Pacific; Onslow, W.A. to central Qld.

Plate 53. Cymatiidae (⁸/₁₀ × natural size)

frog shells (FAMILY BURSIDAE)

THIS RELATIVELY SMALL FAMILY contains about 30 species, all of them living in tropical or warm temperate seas. There is much confusion about the Indo-West Pacific species and about the identity, relationships and number of Australian species in particular. Several generic names are currently in use for members of the family. *Tutufa* appears to be the correct generic name for the species *bubo, bufo* and *rubeta,* while *Colubrellina* may be correctly used for *granularis.* The correct generic names for the other frog shells illustrated are uncertain and for them we have used the traditional name *Bursa* in the broad sense. An older, widely used but invalid name is *Ranella.*

Frog shells are close relatives of the tritons and trumpets from which they may be distinguished by the presence of a deep notch at the posterior end of the aperture representing the posterior canal. The anterior canal is short. Like tritons and trumpets, frog shells have strong varices and a horny operculum. They are usually heavy and strongly nodulose or spiny, and covered with calcareous growths. They take their vernacular name from the warty toad-like appearance of many species. Most frog shells live in shallow water among coral or rocky reefs where their rough sculpture affords them good camouflage. Some species live in sand or mud. Very little is known of their habits. Houbrick and Fretter (1969) reported that *Bursa cruentata, B. rhodostoma* and *Colubrellina granularis* feed on marine worms which they swallow whole. The females lay clutches of stiff, urn-shaped egg capsules on the under-surface of stones or corals. There is a pelagic larval stage.

Reference:

HOUBRICK, JOSEPH R. & FRETTER, VERA (1969). Some aspects of the functional anatomy and biology of *Cymatium* and *Bursa.* Proc. malac. Soc. Lond., **38** : 415-429.

1. *Tutufa bufo* RÖDING, 1798

Whorls with weak spiral ribs, 2 or 3 bearing tubercles, others weak nodules. 4 or 5 ribs swell to form thick ridges where they cross the varices. Interspaces between the ribs patterned with fine oblique striations. Outer lip expanded, scalloped, with 5 or 6 pairs of denticles on its inner margin. Posterior canal deep but short, former posterior canals on the spire not prominent. A thin but wide reflected shield formed over the parietal wall and columella, polished and smooth except for a few teeth or short lirations at the anterior end. White, cream or pink, deep interior white, shield and outer lip margin pink or white, throat of aperture rusty red.

14 cm. Moderately common. Indo-West Pacific; Cape Naturaliste, W.A. to N.S.W. In W.A. this species is commonly taken with hermit crabs in craypots set in 20 to 60 fathoms. *T. bufo* resembles *T. bubo,* but may be distinguished by the smooth parietal shield.

2, 2a. *Tutufa bubo* LINNAEUS, 1758

Whorls with strong spiral ribs, those at the shoulders and at the centre of the body whorl bear large protruding tubercles which may be paired and granulose, other ribs smaller and merely heavily nodulose, tubercles being most prominent where they cross the varices. Outer lip expanded, scalloped, with pairs of denticles on its inner margin. Posterior canal deep but short; former posterior canals on the spire not prominent. Parietal wall calloused, columellar callus raised and reflected to form a shield, columella strongly transversely lirate. White, heavily blotched and spotted with brown, aperture white but usually pink or red in immature shells.

30 cm. Moderately common. Indo-West Pacific; eastern Qld. There has been much confusion about the correct name for this species. One widely used is *lampas* LINNAEUS, 1758. The name *Bursa rubeta gigantea* SMITH, 1914 has also been applied to large specimens from Qld. In fact *rubeta* RÖDING, 1798 is a medium-sized species with a red aperture which also occurs in Qld. The form Smith named *gigantea* does not have close affinity with it but appears to be an Indian Ocean form of *T. bubo.* (2) represents a typical specimen from shallow water in North Qld. (2a) is a specimen trawled off Moreton Bay, Qld. which has a smoother surface than typical *B. bubo* but the columellar shield is similarly rugose. *B. latitudo* GARRARD, 1961 from Tin Can Bay, Qld. is like this but is even smoother and has a higher spire and more angulate shoulders.

3, 3a. *Bursa bufonia* GMELIN, 1791

Very heavy and solid, rather flattened. Sculptured with broad, thick, heavily nodulose spiral ribs. Posterior canal deep, semi-tubular and long. Former posterior canals persist on the spire as long hollow spines. Outer lip expanded, with 4 or 5 pairs of heavy denticles on its inner margin. Parietal wall and columella heavily calloused and coarsely, transversely lirate. Cream or white, profusely spotted with brown, aperture white.

8 cm. Common. Indo-West Pacific; eastern Qld.

4, 4a. *Bursa rosa* PERRY, 1811

Surface with weak granulose cords, 2 spiral ribs which fuse to form 2 or 3 double knob-like protuberances between the varices, and a smaller nodulose spiral rib nearer the anterior end. All 3 ribs thicken as they cross the varices. Outer lip rather narrow, weakly scalloped, with 4 or 5 pairs of small denticles on its inner margin. Posterior canal semi-tubular and very long; former posterior canals

on the spire persist as long hollow spines. Parietal wall and columella thickly calloused but a reflected shield not formed. Strong nodules present on the columella anteriorly, and long but weak irregular lirae centrally and posteriorly on the parietal wall. Exterior cream faintly blotched with purple-brown, deep interior white, throat of aperture coloured deep wine-red.

4 cm. Uncommon. Indo-West Pacific; North-West Cape, W.A. to eastern Qld. *B. siphonata* REEVE, 1844 is an invalid name sometimes used for this species.

5. *Bursa cruentata* SOWERBY, 1835

Whorls with spiral rows of closely and regularly spaced nodules, 3 sets of prominent projecting tubercles at the varices, and 2 strongly tuberculate spiral ribs at the centre of the whorls. Tubercles granulose and double. Outer lip expanded, irregularly scalloped, with 4 or 5 pairs of solid denticles on its inner margin. Posterior canal short, former posterior canals on the spire barely discernible. A thin wide reflected shield present over the parietal wall and columella, the whole shield and columellar margin covered by strong nodules and irregular lirae. Exterior white or cream with a few purple-brown blotches mainly on the tubercles, aperture white, parietal wall usually with conspicuous red-brown spots between the lirae.

4 cm. Common. Indo-West Pacific; eastern Qld. The spots on the parietal wall will usually serve to distinguish this species from immature specimens of *T. bubo.*

6, 6a. *Bursa rana* LINNAEUS, 1758

Slightly flattened, especially at the anterior end. Spire high; 2 narrow elevated fin-like varices lie opposite each other in the horizontal plane on each whorl. Surface sculptured with weak, granulose and nodulose spiral ribs, and prominent pointed tubercles where the ribs cross the varices. Anterior canal up-turned, posterior canal deep, of moderate length and bordered by strong parietal nodules. Strong and crowded denticles present along inner margin of outer lip, and often a long spine at its posterior end. Parietal wall and columella calloused, heavily and irregularly lirate, columellar callus usually raised to form a thin reflected shield. Whitish or pale brown with darker brown spots and spiral bands on the ribs, outer lip and columellar callus white or yellow-brown.

7 cm. Moderately common in mud or muddy sand. Indo-West Pacific; Kimberley, W.A. to eastern Qld. *B. cavitensis* REEVE, 1843 is a variant of this species.

7, 7a, 7b. *Colubrellina (Dulcerana) granularis* RÖDING, 1798

High-spired, rather flattened. Varices strong but low and rounded. Whorls sculptured with weak spiral ribs which bear rounded nodules, and fine granulose spiral striae in the interspaces. Ribs thicken where they cross the varices. Posterior canal wide and short, former posterior canals on the spire not prominent. About 14 strong denticles which tend to be arranged in pairs present on inner margin of outer lip. Parietal wall and columella thickly calloused, transversely lirate. Orange-brown or red-brown with darker spiral lines. Interior, outer lip and columellar callus white.

6 cm. Very common. Indo-West Pacific; Pt. Cloates, W.A. to N.S.W. Many names have been used for this variable species. (7) and (7b) represent the typical form. (7a) is of a specimen in which rows of tubercles replace the granules. This form, named *affinis* BRODERIP, 1832, is regarded as a distinct species by some authorities; *semigranosa* LAMARCK, 1822 is a smooth form. *Colubrellina cubaniana* ORBIGNY, 1842 is a closely related West Indian species.

Plate 54. Bursidae and Colubrariidae (⁹⁄₁₀× natural size)

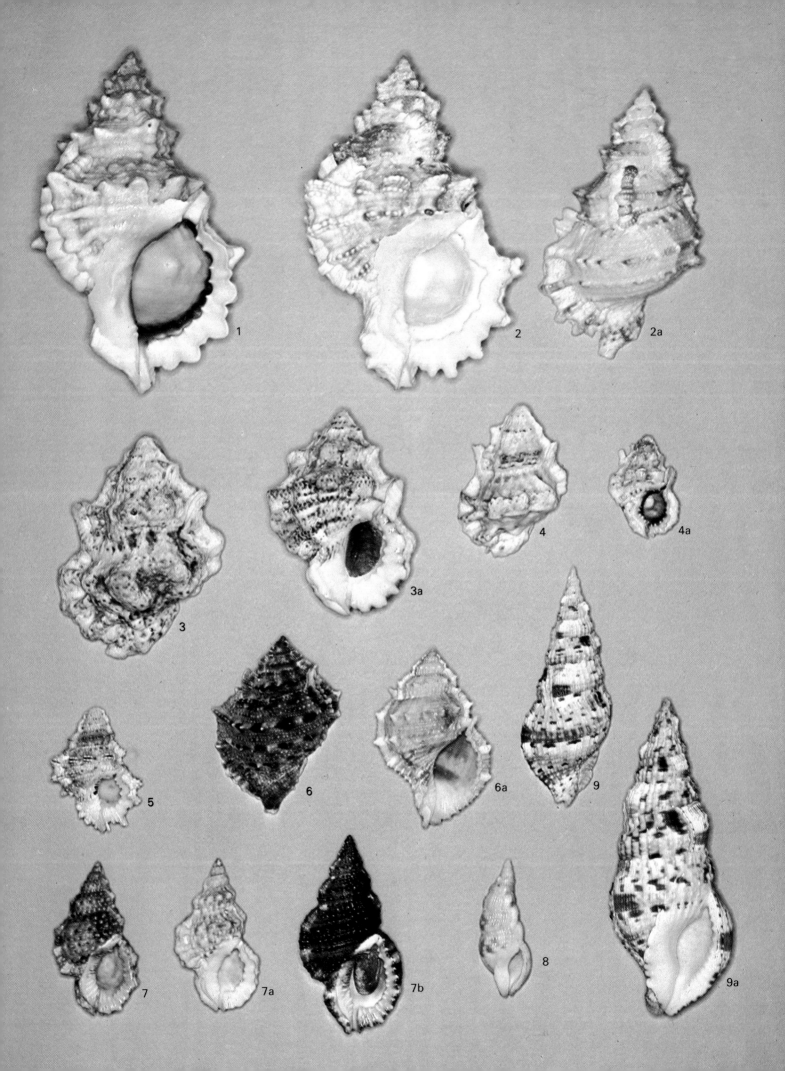

dwarf tritons (FAMILY COLUBRARIIDAE)

ONLY A HANDFUL OF AUSTRALIAN SPECIES can be confidently assigned to the family Colubrariidae. Several genera of small shells sometimes placed here probably belong to other families. Only 2 better known species are illustrated here to represent the family. The colubrariids inhabit warm seas and shallow water. Although at one time they were placed in the family Cymatiidae, they are now thought to be more closely related to the buccinid whelks. Virtually nothing is known of their biology. The shells are long, narrow and fusiform. They have strong varices, granulose or nodulose sculpture and usually a calloused and reflected columella. The anterior canal is short and up-turned, while the posterior canal is represented only by a weak notch at the posterior end of the aperture. The operculum is horny with a terminal nucleus.

Refer to page 81, plate 54, for colour illustrations (8, 9, 9a) of shells.

8. *Colubraria tortuosa* REEVE, 1844
Aperture moderately wide, spire high, pointed and usually twisted. Whorls sculptured with spiral rows of small regularly spaced rounded nodules. Varices smooth except where the rows of nodules cross them. Outer lip with about 15 denticles on its inner margin. Parietal wall thickly calloused, with a strong parietal denticle. Columellar callus thick, reflected and expanded to form a wide shield or anterior rim, columella smooth except for a few weak folds anteriorly. Yellowish or pale brown with wavy brown bands or spots, and large orange-brown blotches on the varices. Interior, outer lip and columella ivory-white and glossy.

5 cm. Uncommon. Indo-West Pacific; Kimberley coast, W.A. to eastern Qld. The smaller size, twisted spire and more regular nodulose sculpture distinguish this species from *C. maculosa*.

9, 9a. *Colubraria maculosa* GMELIN, 1791
Aperture elongate, spire high and pointed, with straight sides. Whorls sculptured with axial rows of small rough nodules and weak spiral striae. Outer lip has 10 to 15 denticles on its inner margin. Parietal wall and columella thickly calloused, columellar callus expanded to form a thick anterior rim, columella smooth or with weak folds. Cream with pale brown spiral bands and 2 or 3 interrupted spiral lines of purple-brown. Interior, outer lip and columella cream, polished.

9 cm. Moderately common. Indo-West Pacific; N.T. to eastern Qld.

murex shells and their relatives

(FAMILY MURICIDAE)

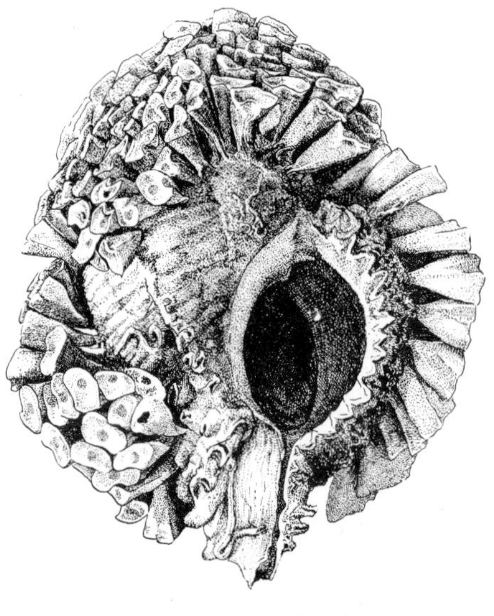

Fig. 17. Egg capsules attached to back of *Hexaplex stainforthi* REEVE, 1842, drawn from a preserved specimen, Port Hedland, W.A.

THIS LARGE FAMILY contains a variety of forms and a number of subfamilies are recognized. Some malacologists treat the thaid whelks as a separate family (Thaididae) but there is little anatomical evidence to support this separation. In this book the thaids are treated as a subfamily only (Thaidinae, pages 90-95).

In his original genus *Murex*, LINNAEUS (1758) placed many species for which new generic names have since been erected, and even some species which are now placed in different families. The generic name *Murex* is now restricted to a group of very spiny shells with long anterior canals (*M. tribulis* L. is the type species) but the term "murex shells" is still used

as a popular name for members of the Muricinae and many members of the other subfamilies.

Murex shells are found in tropical and temperate seas throughout the world. One recent estimate places 1,000 living species in the family.

In both southern and northern Australia the family is well represented by a variety of attractive shells, with as usual, the most striking forms in the tropical north. Nearly all the southern species are endemic to Australia, but the majority of species in the north are also found in other parts of the central Indo-West Pacific region.

Murex shells have a large horny operculum, the spire is usually prominent, the aperture is oval with folds or pointed teeth on the outer lip, and the columella is either smooth or denticulate. The most noteworthy characteristic of murex shells is their strong development of varices, which very often bear prominent and sometimes bizarre spines, nodules or fronds. Few shells excite such immediate interest, and perhaps if they were not so difficult to clean and make presentable for the shell cabinet their popularity among collectors might equal that of the cowries, cones and volutes.

Varical spines and fronds are formed by folds or tucks in the mantle edge. Some of them may be absorbed at later stages of spiral growth so that they do not block the aperture. The function of varical folds in strengthening the shell is readily understood, but the function of spines or fronds seems more open to conjecture. Their most likely function is protection (who would care to swallow a *Murex pecten?*), but in some cases they probably camouflage the shell, aided by adhering growths in older animals.

Members of this family are carnivorous and some of them feed on other molluscs which they attack by drilling holes in the shells. Drilling is accomplished by the joint action of the rotating radula and acid secretions from a special gland below the mouth.

Female murex shells lay tough egg capsules each containing several or many eggs. The form of the capsules varies between species. They may be flask-shaped or dome-shaped, and may be laid separately or close together. Some species are known to congregate for spawning, and lay their capsules on each other's backs (e.g. *Hexaplex stainforthi*). Often there is a brief planktonic larval stage after hatching, but in *Chicoreus territus* (MURRAY and GOLDSMITH, 1963) and *C. denudatus* (Neville Coleman, pers. comm.) development is direct.

The habitats of murex shells also vary. Many live among rocks usually in rather muddy places where their mud-catching spines provide excellent camouflage. Others live under rocks, on coral reefs, or buried in muddy sand among rocks. One species *(Pterynotus permaestus)* lives among mangrove roots.

Selected references:

CERNOHORSKY, W. O. (1967). The Muricidae of Fiji (Mollusca : Gastropoda) Part 1—Subfamilies Muricinae and Tritonaliinae. Veliger 10 : 111-132, pls 14-15.

COTTON, B. C. (1956). Family Muricidae. Roy. Soc. S. Aust., Malacological Section Publ. No. 8, 8 pp., 2 pls.

MURRAY, F. V. & M. H. GOLDSMITH (1963). Some observations on the egg capsules and embryos of *Torvamurex territus* (REEVE, 1845). J. malac. Soc. Aust. 1 (7) : 21-25, pls 2-4.

VOKES, EMILY H. (1964). Supraspecific groups in the subfamilies Muricinae and Tritonaliinae (Gastropoda : Muricidae). Malacologia, 2 : 1-41.

Plate 56. *Pterynotus barclayi* REEVE, 1857.
(Reproduced natural size)

Spire pointed and of medium height, anterior canal moderately long, closed and re-curved. With 3 varices per whorl, each flaring anteriorly to form a wide, thin and fluted frond; 2 to 4 nodulose axial folds between the varices, crossed by rather strong spiral ribs. Inner side of outer lip toothed; parietal wall with a weak nodule beside the posterior canal notch; columella smooth and broad anteriorly.

13.4 cm. Few specimens known. Indo-West Pacific; eastern Qld. The original specimen came from Mauritius; the specimen (above) was recently trawled off Cape Moreton, Qld.

Plate 55. *Haustellum tweedianum* MACPHERSON, 1962, Qld. Close-up study showing foot, eyes and operculum.
(*Photo*—Don Byrne).

1. *Murex nigrospinosus* REEVE, 1845

Body whorl inflated, spire height and aperture length about equal. Anterior canal very long and straight. Varical spines of moderate length tending to be rather straight, the 2 most posterior spines on each varix longer than the others. Whorls sculptured with finely beaded spiral ribs. White or fawn, tips of spines blue or brown. Operculum with fine concentric growth lines.

9 cm. Uncommon. Indo-West Pacific; eastern Qld. Though often misidentified as *M. tribulus*, it lacks the prominent tooth on the lip typical of that species. In Qld. there is another species, of uncertain identity, which is similar to this but has a strongly foliate operculum.

2. *Murex pecten* SOLANDER, 1786

Body whorl inflated, spire height and aperture length about equal. Sutures rather deep, anterior canal very long and straight. Sculpture consists of nodulose spiral ribs with low crescent-shaped axial lamellae in the interspaces. On each varix there is a row of about 17 very long, pointed spines alternating with shorter and thinner intermediate spines, the latter projecting at an angle to the others. Outer lip weakly denticulate, columella smooth, forming a vertical lip anteriorly. Cream or fawn with whitish nodules on the ribs. Operculum with fine concentric growth lines.

13 cm. Uncommon, Indo-West Pacific; eastern Qld. *M. triemis* PERRY, 1811 is a synonym.

3, 3a. *Murex acanthostephes* WATSON, 1883.

Spire of medium height, body whorl inflated, anterior siphon canal very long and usually straight. About 11 long, curved, pointed spines adorn varices on the body whorl and anterior canal. Weak spiral ribs present between the varices and in the interspaces there are tiny spirally-elongate "blisters" which are open at the ends facing the lip. Fragile axial lamellae present in the sutures. Outer lip with about 6 strong teeth, one near the anterior end prominent and projecting beyond the others. Spire nucleus sharply keeled and tan coloured. Exterior fawn or cream with white flecks, tips of the larger spines dark grey, aperture white.

10 cm. Moderately common. North-West Cape, W.A. to Torres Strait, Qld. The original specimens were dredged from 28 fathoms in the Arafura Sea and have longer spines than the specimens illustrated here which came from shallow water at Port Samson, W.A.

4. *Murex coppingeri* E. A. SMITH, 1884

Stout, with a rather high shouldered spire and a relatively short tapered anterior canal. Surface sculptured with fine but sharp and distinct spiral cords, a heavy angular rib at the shoulder, and 1 or 2 axial nodules between each pair of varices. Varices bear 7 major spines, the most posterior longest and strongly curved. Short spinelets present between the major spines. Outer lip margin weakly toothed. Fawn with light brown spiral lines.

5 cm. Uncommon. The only specimens we have seen are from Broome, W.A., Darwin, N.T. and the Gulf of Carpentaria, Qld.

5. *Murex macgillivrayi* DOHRN, 1862

Rather solid, with a moderately high spire and a long, narrow, straight anterior canal. Each varix bears 4 short, curved, pointed spines. Between the varices there are 2 axial rows of nodules and well spaced, fine spiral cords. Outer lip bears flattened teeth, columella smooth, forms a vertical lip anteriorly. Creamy-white to fawn with brown cords and a brown patch at the anterior extremity.

8 cm. Uncommon. Central Indo-West Pacific; North-West Cape, W.A. to eastern Qld. Whether this species should be in the genus *Haustellum* or in the genus *Murex* is uncertain.

6, 6a. *Haustellum tweedianum* MACPHERSON, 1962
(see also colour plate of live animal, page 83, plate 55)

Solid, with a moderately high spire, and a long narrow straight anterior canal. Each varix crossed by low beaded ribs and cords which do not bear spines, although a few spines are present at the base of the anterior canal. 2 axial rows of nodules present between the varices. Outer lip toothed, columella smooth, forms a vertical lip anteriorly. Creamy-white with patches of brown and sometimes of pink, spire pink.

8 cm. Moderately common. Trawled off southern Qld. and N.S.W.

7, 7a. *Haustellum multiplicatum* SOWERBY, 1895

Spire rather low, body whorl short, anterior canal a long thin straight spike. 5 axial folds present between the thick rounded varices, and about 14 spiral cords on the body whorl, forming nodules where they cross varices and folds. Varices continue along the anterior canal spike. Columella thin, forms a high vertical lip, outer lip weakly toothed. Pale fawn with broad brown spiral bands, the most anterior extending on to the anterior canal. Beads on the varices and plications white or pale fawn, anterior canal light brown, interior white.

7 cm. Uncommon. North-West Cape, W.A. to the N.T. Some specimens have a few short spines on the varices.

8. *Pterynotus (Naquetia) permaestus* HEDLEY, 1915

Solid; compact, with a high spire, spirally ribbed whorls and a broad short anterior canal. 3 varices per whorl, each varix ribbed and frilled but spineless, sometimes flared anteriorly. Outer lip toothed; columella smooth. Chocolate-brown; aperture light brown or grey.

6.5 cm. Common. Central Indo-West Pacific; North-West Cape, W.A. to Qld. Lives among roots of mangrove trees (see Vokes, 1964, pp.15-16).

9, 9a. *Hexaplex stainforthi* REEVE, 1842

Solid, ovate, with a rather low spire and a broad, short anterior canal. 8 to 9 varices per whorl, each bearing numerous short delicate fronds. Whorls sculptured with coarse spiral ribs running up on to the fronds. Outer lip toothed, columella smooth. Deep umbilicus. White, pale pink, orange or yellow spiral ribs, fronds dark brown, aperture pink, orange or yellow.

6.5 cm. Common. North-West Cape, W.A. to N.T. This species is usually placed in the genus *Bassiella* WENZ, 1941 but Vokes (1964) places that name into synonymy with *Hexaplex* PERRY, 1810.

10. *Homalocantha anatomica* PERRY, 1811

Spire high, anterior canal long and straight, whorls rather angulate. Sutures very deeply excavated and crossed by thick axial walls. With 5 or 6 varices per whorl, each bearing long digitations expanded at their ends. Usually 2, sometimes 3 major digitations on that part of the varix which borders the aperture. Whorls smooth between the varices except for prominent spiral cords corresponding to the major digitations. Aperture oval and smooth, but there may be small spurs along the ventral surface of the last varix. White.

6 cm. Uncommon. Indo-West Pacific; eastern Qld. In this remarkable shell the numbers of varical digitations are variable and several forms of the species have been named. Although the matter is not finally settled it now seems likely that *M. rota* MAWE, 1823 and *M. pele* PILSBRY, 1921 are synonyms of *H. anatomica*. More details of the variations found in Australian waters are needed.

11, 11a. *Homalocantha secunda* LAMARCK, 1822

Solid, flattened, with a rather short spire and a broad, flat and rather long anterior canal. With 6 varices per whorl, the last broad and bearing 8 to 10 flat broad scaled fronds, each slightly wider at its distal end. Whorls coarsely ribbed between the varices. Sutures deeply excavate and crossed at the varices by smooth thin lamellae connecting the whorls. Outer lip edge rough but not toothed, aperture almost round. Fawn or light brown, fronds dark brown, interior white.

4.5 cm. Uncommon. Central Indo-West Pacific; North-West Cape to Broome, W.A. *H. varicosa* SOWERBY, 1841 from Qld. may be the same species.

Plate 57. Muricidae ($1\frac{1}{10}$ × natural size)

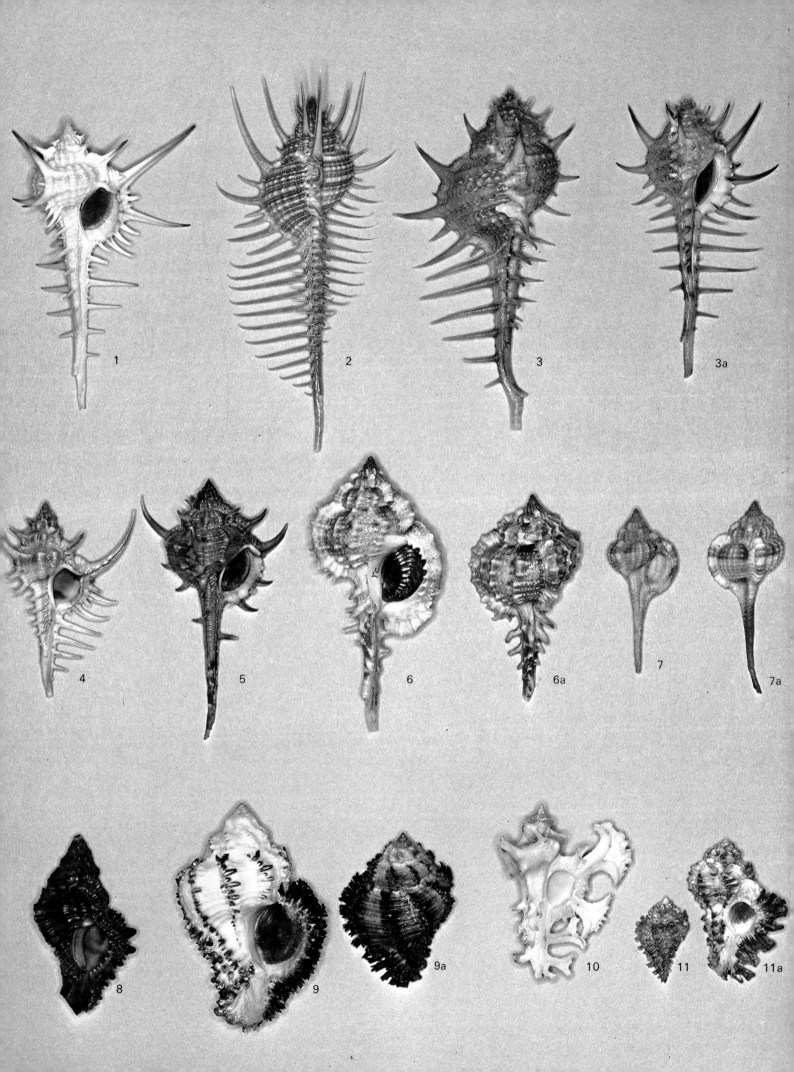

Muricidae (Cont.)

1. *Chicoreus ramosus* LINNAEUS, 1758
Large and solid with a tumid body whorl, moderately high spire, and a long re-curved anterior canal. Varices bear rather short stout up-turned fronds, with 1 or 2 prominent, axially-elongate nodules between them. Surface sculptured with weak spiral ribs and numerous fine spiral cords. Conical pointed teeth present on outer lips including a large one near the anterior end, columella smooth. Exterior white, aperture margins pink or white.
   30 cm. Moderately common. Indo-West Pacific; Broome, W.A. to eastern Qld. Thi is the largest murex of the region. The illustrated specimen is juvenile.

2. *Chicoreus banksii* SOWERBY, 1840
Rather delicate, with a high spire. Anterior canal moderately long, and either straight or slightly re-curved. Each varix bears 8 or 9 broad, long, finely-scaled fronds of which the bifurcate posterior one is the largest. Outer lip toothed, columella smooth. Between each pair of varices there are 1 or 2 rough axially-elongate nodules, and also numerous finely-scaled spiral cords. Exterior light brown with dark brown varices, aperture and columella white.
   8 cm. Uncommon. Indo-West Pacific; northern W.A. The illustrated specimen is from Broome. A similar shell trawled off the north eastern coast of Qld. is known as *C. axicornis* LAMARCK, 1822. At Broome *C. banksii* is found alongside the more common *C. brunneus* LINK 1807 (not illustrated) and the species tentatively identified here as *C. rubiginosus* (4, 4a, 4b).

3. *Chicoreus torrefactus* SOWERBY, 1841
Usually rather slender, with a high spire but a moderately short and slightly curved anterior canal. Short, open bifurcated fronds present on the thick rounded varices on each whorl. Between the varices 1 or 2 rounded axial ridges or elongate nodules present, crossed by numerous strong spiral ribs and fine scaly striae. Outer lip toothed, columella smooth, with 1 or several small parietal denticles. Usual external colour brown, with dark brown spiral ribs and fronds. Aperture and columella straw-coloured or orange.
   10 cm. Common. Indo-West Pacific; North-West Cape, W.A. to eastern Qld. The more slender shell, higher spire, lower varices and characteristic aperture colour distinguish this species from *C. brunneus* LINK, 1807, another common and widespread species with which it is often confused. In *C. brunneus* the aperture and columella are dark rose. (3) is a typical *C. torrefactus* from North-West Cape, W.A. Compare this species with *C. rubiginosus*.

4, 4a, 4b. *Chicoreus rubiginosus* REEVE, 1845
Like *C. torrefactus* but rather more stout, with longer bifurcate fronds. Most specimens brown with darker spiral ribs and fronds (e.g. 4), but sometimes cream (4a), orange (4b), or mauve.
   9 cm. Common. Broome, W.A. This name is suggested only tentatively for the specimens illustrated here, all of which come from Broome. These shells are generally identified as *C. torrefactus* but differ from the typical form of that species in having long varical fronds. If further studies show the Broome shells to be a distinct species, then *C. rubiginosus* is probably the name that should be used for them. Some collectors have been calling these shells *C. steeriae* REEVE, 1845, but that is a different species from Polynesia.

5. *Chicoreus cornucervi* RÖDING, 1798
Rather thin, with a high spire, deeply indented sutures, and a long re-curved anterior canal. Varices bear very long, strongly re-curved widely separated, almost tubular fronds, which carry only small side branches. Scaly spiral ribs and folds present between the varices but no major nodules. Outer lip with several small teeth and one prominent tooth near the anterior end. Columella smooth, posterior canal a wide notch at the posterior end of the lip. Dark or light brown with darker brown varices, apertural margins usually pink. Occasionally cream-coloured with a yellowish aperture.
   11 cm. Common at some localities. North-West Cape, W.A. to the N.T. Once commonly known as *Murex monodon* SOWERBY, 1825.

6, 6a. *Chicoreus damicornis* HEDLEY, 1903
Rather thin, with a high spire and a moderately long, re-curved, open anterior canal. Varical fronds long, broad, open and rather fragile, the most posterior fronds on the shoulders, are the largest and are bifid. One weak nodule present between each pair of varices, surface of whorls irregularly spirally striate. Weak irregular teeth on outer lip, columella smooth. Exterior cream, tinged with pink or brown.
   6 cm. Moderately uncommon. South-eastern Australia between N.S.W. and S.A.

7, 7a. *Chicoreus denudatus* PERRY, 1811
Solid but delicately sculptured, with a long spire and a short, re-curved anterior canal. Varices bear numerous crowded, short and delicately beaded fronds. Two sets of nodules present between each pair of varices. Surface finely, spirally striate, outer lip finely toothed, columella smooth. Exterior dark or light brown, yellow or orange, aperture white.
   5 cm. Moderately common. South-eastern Australia from N.S.W. to S.A. This and other species are customarily put in the genus *Torvamurex* IREDALE, 1936, but Vokes (1964) considers that name to be a synonym of *Chicoreus*. The N.S.W. form is sometimes known as *C. extraneus* IREDALE, 1936, but the differences between it and the S.A. and Vic. shells are minor and inconsistent.

8. *Chicoreus recticornis* VON MARTENS, 1880
Spire rather high, anterior canal very long, narrow and straight or slightly re-curved. Each varix bears 2 long, narrow, solid, almost straight spines or spikes which end in sharp points, with 3 or 4 much smaller spines between them. Small spines also present on the anterior canal. Several axial rows of nodules and fine spiral cords present between the varices. Exterior creamy-white or fawn, tips of the spikes brown, aperture white.
   6 cm. Uncommon. Trawled off south-eastern Qld. The shell resembles *C. cervicornis* but the spikes are not branched.

9. *Chicoreus cervicornis* LAMARCK, 1822
Rather delicate, with a moderately high spire and a very long, curved or almost straight anterior canal. Each varix bears 3 or 4 long spines of which the anterior 2 are forked. Additional spines present on the anterior canal. Whorls finely ribbed between the varices. Exterior and aperture creamy-white or fawn.
   6 cm. Uncommon. Rottnest Is. W.A. to northern Qld. The Rottnest Is. record is based on specimens dredged recently at depths of 80 to 100 fathoms.

10, 10a. *Chicoreus huttoniae* WRIGHT, 1879
Rather small, with a high spire and a broad, rather short and almost straight anterior canal. Low varices consist of short, broad, open, frilled fronds. Between them are 3 axial ridges crossed by several spiral riblets and numerous fine scaled lirae. Outer lip toothed, columella smooth. Exterior pink, mauve, white or brown, interior white.
   3.5 cm. Uncommon. New Caledonia and eastern Qld.

11. *Chicoreus territus* REEVE, 1845
Characterized by a high spire, a moderately long, slightly re-curved and open anterior canal, and expanded varices profusely frilled at their margins. These broad up-turned frills alternate with smaller, straighter points. 1 or 2 rough elongate nodules present between the varices, surface crossed by fine spiral cords. Outer lip toothed, columella smooth. Colour variable, exterior white, cream, grey or brown, aperture white.
   7 cm. Not common. Indo-West Pacific; trawled off eastern Qld.

Plate 58. Muricidae (⁸⁄₁₀ x natural size)

Muricidae (Cont.)

1, 1a, 1b. *Pterynotus (Pterochelus) triformis* REEVE, 1845

Rather solid with a high spire, a long, almost straight, open anterior canal, and 3 high thin frilled and fin-like varices per whorl. Posterior canal well developed, forming a back-projecting corner at the posterior end of the last varix. Whorls sculptured with irregular spiral cords and 1 prominent nodule between each pair of varices. Outer lip irregular and sometimes toothed. Exterior brown, yellow or creamy-white with brown patches, interior white.

6 cm. Common on sand among sea grasses. N.S.W. to Albany, W.A. See remarks on *P. acanthopterus*.

2, 2a, 2b, 2c, 2d. *Pterynotus (Pterochelus) angasi* CROSSE, 1863

Small and delicate with a high spire, a broad, moderately long, almost closed anterior canal, and broad, thin, finely fluted and fin-like varices. Posterior canals well developed, forming backward projecting, re-curved, almost closed tubercles at the posterior ends of the varices. Fine spiral striae and several nodules present between each pair of varices. Outer lip smooth. Colour variable, orange, yellow, fawn or brown. Some specimens banded.

2 cm. Moderately common under stones in shallow water, but hard to find. N.S.W. to Fremantle, W.A.

3, 3a. *Pterynotus (Pterochelus) bednalli* BRAZIER, 1877

Spire high, anterior canal moderately long, broad, closed and re-curved. Varices wide, thin, gently fluted and fin-like. Whorls almost glossy, sculptured with about 25 fine spiral riblets which run up the sides of the varical fins. Posterior canal or tube lacking, outer lip toothed. Exterior creamy-white with rusty brown patches, interior white.

8.5 cm. Uncommon. Found on mud flats. The species has a restricted range in the extreme north of W.A. and the N.T.

4, 4a, 4b. *Pterynotus (Pterochelus) acanthopterus* LAMARCK, 1816

Very like *P. triformis* but the backward-projecting corners on the varices formed by the posterior canals more prominent and supported by strong curved ribs.

10 cm. Moderately common. Typical *P. acanthopterus* is found in northern W.A. from North-West Cape to the Kimberley region. Along the central west coast of W.A. between Cape Leeuwin and North-West Cape forms between *triformis* and *acanthopterus* occur, and it is possible that these "species" are extreme end-of-range forms of a single clinal species. The eastern *P. duffusi* is also a member of this group, and some authors believe it to be only a form of *P. acanthopterus*. A further complication is that 2 distinct size forms of *P. acanthopterus* occur in the north of W.A. Large shells up to 10 cm long (4, 4b) are found in the same places as small shells only about 4 cm long (4a). It has been suggested that this is a case of sexual dimorphism, but observations on living specimens are needed to confirm this.

5, 5a. *Pterynotus (Naquetia) triqueter* BORN, 1778

Slender, solid but delicately sculptured. Spire high, anterior canal broad, long and rather straight with a turned-up end. Varices rather thick and rounded posteriorly, but flaring to wide thin fluted fins along the sides of the anterior canal. Posterior canal absent. Between each pair of varices are 2 or 3 nodulose axial ribs, crossed by numerous spiral ribs. Outer lip toothed, columella smooth. Creamy-white with light brown patches, spiral lines or bands.

7 cm. Uncommon. Indo-West Pacific; eastern Qld.

6. *Pterynotus patagiatus* HEDLEY, 1912

Solid, with a high spire and a moderately short but strongly re-curved anterior canal. (The former anterior canals may remain as in the illustrated specimen). Varices foliate, oblique and fin-like, crossed by strong granulose spiral ribs, with 1 large but low nodule between the varices. Outer lip toothed, columella smooth except for 1 posterior nodule, no posterior canal. Varices whitish, whorls between them pale pink, orange-pink or orange-yellow. Tips of the anterior canals brown.

6 cm. Uncommon. Trawled, central Qld. to N.S.W. The holotype and other N.S.W. shells have narrow weakly-foliated varices. The figured specimen is from Qld.

7. *Pterynotus (Naquetia) tripterus* BORN, 1778

Heavy, solid, with a moderately high spire and a broad moderately long and straight anterior canal. Varices wide, thin and fin-like, and covered with fine crescent-shaped lamellae. Surface granulose and sculptured with fine spiral ridges. 6 to 8 denticles present on outer lip, and 7 to 10 on columella. White or fawn.

6 cm. Uncommon. Central Indo-West Pacific; eastern Qld.

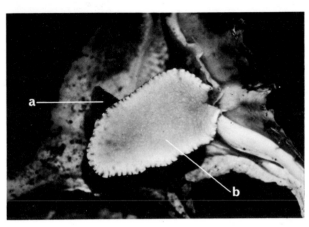

Fig. 18. *Pterynotus (Pterochelus) duffusi* IREDALE, 1936, alive, showing foot (b) and operculum (a). (*Photo*—Neville Coleman).

8. *Pterynotus (Pterochelus) duffusi* IREDALE, 1936

Resembles *P. acanthopterus* of northern W.A., but the posterior canals even more prominent. 2 additional spiral ribs present at the centre of the varices.

6 cm. Uncommon. N.S.W. See *P. acanthopterus*.

9, 9a. *Murexsul umbilicatus* TENISON WOODS, 1875

Solid, with a high spire, up-turned anterior canal, and about 8 low varices on the body whorl, crossed by strong, scaled spiral ribs. Anterior fasciole strong and raised, umbilicus narrow but deep. Outer lip weakly toothed, columella smooth. External colour yellow, slightly darker on the varices, aperture white.

3 cm. Moderately common in shallow water. Vic., Tas., and S.A. *M. brazieri* ANGAS, 1877 and *M. fimbriatus* LAMARCK, 1822 are 2 smaller southern Australian species.

10, 10a. *Typhis philippensis* WATSON, 1886

Oblong-fusiform with a moderately high, shouldered spire, long, narrow, slightly up-turned and tubular anterior canal. Body whorl with 4 spinose varices and between each pair a nodulose axial varix-like ridge which bears a high hollow tube at its posterior end, the last of these tubes occupied in life by the animal's posterior siphon. The round aperture encircled by a smooth raised rim. Exterior shiny and fawn with whitish blotches on the varices and ridges, aperture white.

2.5 cm. Uncommon in collections. Dredged between 10 and 100 fathoms. N.S.W. to Vic. and S.A. *T. yatesi* CROSSE, 1865 is a similar southern Australian species which has broad foliated varices. These 2 species belong in the subfamily Typhinae.

11. *Vitularia miliaris* GMELIN, 1791

Biconic, umbilicate, with a moderately high spire, a tapered almost closed anterior canal, and a smooth columella. With 8 axial-oblique varices on the body whorl, joined by 3 broad low spiral ribs and thin lamellae which are highest near the sutures. Exterior buff or grey, interior white, columella yellow.

5 cm. Uncommon. Indo-West Pacific; eastern Qld.

12. *Bedeva hanleyi* ANGAS, 1867

Biconic to fusiform with a high, shouldered spire and short anterior canal. Sculptured with strong axial folds (about 8 on the body whorl), crossed by fine spiral ribs. Outer lip toothed, columella smooth. Brown or grey, sometimes with yellow or white spiral lines.

2 cm. Common. On sand or mud, or on mussel or oyster beds. Southern Australia from central Qld. to Shark Bay, W.A.

Plate 59. Muricidae (natural size)

Muricidae (Cont.)

purples and drupes (SUBFAMILY THAIDINAE)

MANY HUNDREDS OF SPECIES belong in this subfamily which some malacologists regard as a distinct family. In the tropical Indo-West Pacific region there are a number of medium-sized whelks and a host of small variable species which differ mainly in details of sculpture. Only a few of the better known species are illustrated here. Almost all thaids live among rocks or corals, many of them in the intertidal zone. These carnivorous snails feed mainly on barnacles, bivalves and other molluscs. Some of them are able to drill holes through the shells of their prey. Thaids usually have thick and solid shells which are strongly sculptured with spines, nodules or spinal ribs, but true varices are lacking. The outer lip may be strongly toothed. The anterior siphon canal may be merely a deep notch in the lip or it may be a short channelled snout. The horny operculum has a lateral nucleus.

Selected references:

CERNOHORSKY, W. O. (1969). The Muricidae of Fiji, Part II—Subfamily Thaidinae, Veliger, 11 : 293-315, pls 47-49.
MAES, VIRGINIA ORR. (1967). The littoral marine fauna of Cocos-Keeling Islands (Indian Ocean). Proc. Acad. Nat. Sci. Philadelphia, 119 : 93-217.

1. *Thais bufo* LAMARCK, 1822
Broadly ovate, spire of low to medium height. Body whorl with numerous incised spiral lines which end in denticles on the edge of the outer lip, 4 spiral rows of low nodules, and raised anterior and posterior fascioles. Aperture smooth. Columella calloused and smooth, posterior canal well formed and bordered by a thick parietal fold. Exterior dark brown with thin white spiral lines and white patches between the nodules, lip margin dark brown, deep interior cream becoming yellow near the lip margin and on the polished parietal wall and columella.
2.5 cm. Moderately common. Indo-West Pacific; N.T. to eastern Qld.

2, 2a. *Mancinella mancinella* LINNAEUS, 1758
Broadly ovate, spire moderately high. Body whorl with many fine spiral striae, 5 spiral rows of pointed nodules, and a thick, scaled anterior fasciole. Aperture outer wall with fine spiral lirae. Posterior canal obsolete. Exterior creamy-white, pale rose or orange-red, with darker red-purple on the nodules. Aperture, parietal wall and columella yellow to dark orange or orange-red, internal lirae darker orange-red.
5 cm. Common. Indo-West Pacific; Dorre Is., W.A. to northern N.S.W.

3. *Mancinella tuberosa* RÖDING, 1798
Biconic, spire moderately high. Body whorl with fine spiral cords which end in minute denticles at the edge of the outer lip, 2 posterior spiral rows of solid, conical tubercles, 1 spiral row of low nodules, and a thick rough anterior fasciole. Aperture with numerous fine spiral lirae on the outer wall and 2 weak folds on the columella. Posterior canal obsolete, bordered by a weak parietal fold. Exterior black or purple-brown with white patches between the nodules. Aperture white or cream, lirae yellow or pale orange, columella and parietal wall polished, cream with purplish or brown blotches.
5 cm. Common. Indo-West Pacific; Broome, W.A. to eastern Qld. The species is put in the genus *Mancinella* rather than in *Thais* because of radular characteristics. (Cernohorsky, 1969). *T. pica* BLAINVILLE, 1832 is a synonym.

4. *Thais armigera* LINK, 1807
Biconic, spire high. Body whorl with weak spiral striae which end in tiny teeth at the edge of the lip, 3 spiral rows of prominent blunt spines (the most posterior high and projecting), and a high, rough, angular anterior fasciole. Aperture with a series of small denticles on the inner margin of outer lip, 3 or 4 weak lirae on the columella and an obsolete parietal denticle. Posterior canal obsolete. Spines and fasciole white or fawn, usually eroded, interspaces with brown spiral bands, deep interior white, becoming yellow near the lip, parietal wall and columella polished, brown or orange.
8 cm. Common. Intertidal reefs exposed to surf. Indo-West Pacific; eastern Qld.

5, 5a. *Thais aculeata* DESHAYES, 1844
Ovate to biconic, spire moderately high. Body whorl with fine spiral grooves which end in minute pointed denticles on the outer lip, 4 spiral rows of stout conical nodules (the nodules of the 2 posterior ribs are largest) and a thick anterior fasciole. Aperture with 5 lirae terminating at medium sized denticles on the inner margin of the outer lip. Columella with 1 or more weak folds. Posterior canal a shallow notch. Exterior black or dark purple-brown with white patches or streaks between the nodules, deep interior purplish or bluish white, inner edge of outer lip black or purple-brown, parietal wall and columella black, brown or purple-brown, with white patches.

6 cm. Common. High in the intertidal zone of rocky shores. Indo-West Pacific; Pt. Cloates, W.A. to eastern Qld. This is *T. hippocastanum* of authors *non* LINNAEUS, 1758.

6, 6a. *Thais kieneri* DESHAYES, 1844
Biconic, spire rather high, sometimes equal to ½ the total shell length. Body whorl with strongly incised spiral lines which end in tiny teeth at the edge of the lip, 4 spiral rows of nodules (those of the most posterior row being high, pointed and compressed), and a low, rough anterior fasciole. Aperture and columella smooth, outer lip thickened in adults and bears 4 to 5 rather large denticles on its inner margin. Posterior canal a shallow notch. Exterior dark brown with white or yellow spots, blotches and axial streaks between the nodules, 4 or 5 brown patches on the lip margin. Deep interior white or cream becoming yellow at the lip and on the polished columella and parietal wall.
6 cm. Common. High in the interidal zone on rocky shores. Indo-West Pacific; North-West Cape, W.A. to eastern Qld. It resembles *M. tuberosa* but the spire is high, the columella lacks the brown blotches and the aperture lacks the spiral lirae of that species.

7, 7a. *Thais echinata* BLAINVILLE, 1832
Biconic, sometimes umbilicate, spire moderately high. Body whorl with fine spiral striae, 4 spiral rows of pointed nodules, and a thick, elevated, scaled anterior fasciole. Aperture smooth. Posterior canal obsolete. Creamy-grey to light brown externally, aperture, parietal wall and columella white.
4 cm. Common. Intertidal coral reefs. Indo-West Pacific; Shark Bay, W.A. to eastern Qld.

8. *"Dicathais" baileyana* TENISON WOODS, 1881
Biconic, spire height about ⅓ of the total length. Anterior canal deep, anterior end attenuate. Whorls rounded and sculptured with numerous concentric ribs, fine undulating axial lamellae, and a few irregular axial ribs representing stages of growth. Aperture wall spirally lirate, outer lip crenulate. Columella smooth, angled at the base of the anterior siphon canal. Exterior whitish or creamy-grey, interior and columella white.
3.5 cm. Uncommon. Intertidal rocky shores. Vic., Tas. and S.A. The generic placement of this species is questionable. The shell may be distinguished from *D. textilosa* by the finer and more regular spiral ribs and the attenuate anterior end.

9, 9a. *Dicathais textilosa* LAMARCK, 1822
Ovate, spire low or of medium height. Variously sculptured, usually with 8 or 9 spiral ribs, more numerous fine spiral cords, and fine irregular axial growth lamellae. Ribs sometimes nodulose. Anterior fasciole broad and low. Inner margin of outer lip spirally grooved. Posterior canal a shallow notch bordered by a weak parietal denticle. Exterior cream or grey, aperture white becoming yellow near the outer lip, anterior end of the columella yellow.
8 cm. Common. Among rocks in the intertidal and sublittoral zones. Vic. and S.A. to Barrow Is., W.A. Shells from W.A. (9) are often heavily nodulose and have been named *D. aegrota* REEVE, 1846. (9a) represents a typical *"textilosa"* shell from Vic.

10. *Dicathais orbita* GMELIN, 1791
Another member of the *textilosa-aegrota* complex. Characterized by 8 or 9 high, broad, spiral ribs with wide, deep grooves between. Five irregular cords present on the ribs and in the grooves, small ribs sometimes also present in the grooves.
8 cm. Common. Among rocks in the intertidal and sublittoral zones. Southern Qld. to eastern Vic. Intergrades between the typical *D. orbita* and *D. textilosa* forms occur and they may not be specifically distinct. *D. orbita* has much stronger spiral sculpture and the spire tends to be higher than in *D. textilosa*. *D. scalaris* MENKE, 1829 is another similar form from New Zealand.

Plate 60. Thaidinae (1¹⁄₁₀ × natural size)

1, 1a. *Drupa rubusidaeus* RÖDING, 1798

Ovate, humped, spire rather low. Body whorl with 5 rows of open-pointed nodules or spines, a high anterior fasciole which bears long open spines, and in the interspaces numerous fine spiral scaly cords which end in fine denticles at the edge of the outer lip. Inner margin of the outer lip has about 12 small even-sized denticles. Anterior end of the columella with 3 to 5 plicate folds. Basal margin of the anterior left side strongly flanged and crenulate. Exterior yellow, aperture lavender, yellow at the margins.

6 cm. Moderately common. Intertidal reefs. Indo-West Pacific; Qld. Note the hoof limpet (*Hipponyx* sp.) attached to the side of the specimen illustrated from the ventral aspect.

2. *Drupa morum* RÖDING, 1798

Flat-based, with a low spire. Body whorl with 4 spiral rows of large conical nodules, an anterior spiral rib with lower nodules, and a spinose anterior fasciole. Fine scaly spiral cords present between the rows of nodules. Outer lip finely denticulate along its edge, inner margin with 2 separate denticles anteriorly, a fused pair of denticles centrally, and 4 fused denticles posteriorly. Columella with 3 or 4 strong plicate folds. Anterior canal short and narrow, posterior canal a long shallow groove. Exterior white, nodules brown or black, aperture dark violet.

4 cm. Common. Intertidal reefs. Indo-West Pacific; Barrow Is., W.A. to eastern Qld.

3, 3a. *Drupa ricinus* LINNAEUS, 1758

Ovate, flat-based, spire low. Body whorl sculptured by fine spiral striae, 4 spiral rows of long spines and a spinose anterior fasciole. Inner margin of outer lip with 2 small separate denticles anteriorly, 2 fused thick denticles centrally, and a block of 4 fused denticles posteriorly. Columella with a pair of fused denticles. Exterior white, spines black or black-tipped; aperture white, often with yellow or orange blotches around margins.

3 cm. Common. Intertidal reefs. Indo-West Pacific and as far east as the Clipperton and Galapagos Is., Houtman Abrolhos, W.A. to eastern Qld. Some authors reserve this name for a form with an entirely white aperture (3), and call the form with orange or yellow blotching *D. arachnoides* LAMARCK, 1816 (3a), but Cernohorsky (1969) has demonstrated that they are colour forms of a single species.

4. *Drupina lobata* BLAINVILLE, 1832

Like *D. grossularia* in most respects. Spire higher in *D. lobata,* with exterior black, columella parietal wall and inner margin of outer lip glossy black, deep interior pale yellow.

3 cm. Uncommon. On intertidal, wave-swept reefs. Indian Ocean; recently recorded from Pt. Cloates, W.A. Some authors regard this as a colour form or subspecies of *D. grossularia,* but the colour difference is quite distinctive, and the presence of both forms in W.A. suggest to us that they should be regarded as distinct species.

5, 5a. *Drupina grossularia* RÖDING, 1798

Low, flat-based. Spire low, skewed. Exterior scaly, with 4 nodulose spiral ribs and a strongly scaled anterior fasciole. Outer lip with 3 short pointed projections anteriorly and 2 longer, flat bifurcated digitations posteriorly, the most posterior one open and functions as the posterior siphon canal. Inner margin of outer lip with 5 or 6 denticles. Columella calloused, with 2 or 3 small denticles anteriorly. Exterior grey, usually heavily overgrown with calcareous growths, aperture and columella yellow.

4 cm. Common. On intertidal wave-swept reefs. Indo-West Pacific; Barrow Is., W.A. to Qld.

6, 6a. *Morula spinosa* H. & A. ADAMS, 1853

Elongately biconic, spire high or of medium height, anterior canal long and projecting. Body whorl sculptured with 3 spiral rows of long spines aligned on low axial varices, a row of small spines anteriorly, a heavily-scaled fasciole, and numerous fine spiral cords in the interspaces. Penultimate whorl with 1 row of spines. Outer lip with 5 or 6 strong denticles on inner margin. Columella reflected forming a shield, with 5 or 6 small plicae anteriorly. Colour variable, exterior usually cream or pale brown with brown spire and varices, aperture and columella usually violet.

3.5 cm. Moderately common. Coral reefs in shallow water. Indo-West Pacific; Houtman Abrolhos, W.A. to northern N.S.W.

7, 7a. *Morula uva* RÖDING, 1798

Ovate, spire of medium height. Sculptured with spiral rows of pointed tubercles, 5 rows on the body whorl and 2 on the spire whorls, with scaly grooves in the interspaces and a short, rough, tightly re-curved anterior fasciole. Outer lip with 2 large peg-like denticles near the centre, and as many as 3 smaller denticles anteriorly. Columella thickened, with 3 or 4 small denticles or plicae. Exterior white with black tubercles and anterior fasciole, aperture and columella violet.

3 cm. Common. Intertidal and shallow sub-littoral rock and coral reefs. Indo-West Pacific; Houtman Abrolhos, W.A. to northern N.S.W.

8. *Morula marginalba* BLAINVILLE, 1832

Biconic with 5 spiral rows of heavy rounded nodules on the body whorl, plus a short nodulose anterior fasciole. Interspaces finely spirally striate between the nodules. Spire height greater than aperture length, anterior canal short, posterior canal notch shallow. Aperture with 4 spiral lirae which end (in adults) at 4 prominent evenly spaced denticles on the inner margin of the lip. 2 or 3 weak plicae sometimes present at the anterior end of the otherwise smooth, rather straight columella. Yellow-grey ground colour with black to dark purple nodules and anterior fasciole, outer lip yellowish with 4 black patches in immature shells, columella white, deep interior blue-white.

3 cm. Abundant on rocks high in the intertidal zone, often among oysters. Central Qld. to N.S.W. This is the "Mulberry Whelk" reputed to damage oyster beds. It is a close relative of the common Indo-West Pacific species *M. granulata* DUCLOS, 1832 which has a smaller aperture and tends to have the two posterior denticles of the outer lip closer together.

9. *Morula margariticola* BRODERIP, 1832

Biconic, spire height slightly less than total shell length. Body whorl with about 9 broad axial folds, terminating in prominent nodes at the shoulders, and crossed by numerous scaly spiral cords. Anterior fasciole strong scaled. Anterior canal short. Outer lip with 3 or 4 small denticles on its inner margin. Columella smooth except for a few very fine but elongate anterior plicae. Exterior black or dark brown, whitish in the grooves between the spiral cords, deep interior pale blue-white, inner margin of outer lip dark brown or black, columella brown.

3 cm. Abundant on rocks in high intertidal zone, often among oysters. Indo-West Pacific; North-West Cape, W.A. to eastern Qld. The shape of the shell of this species is quite variable. Some specimens, like the one figured, are broadly bi-conic, others are more elongate.

10, 10a. *Lepsiella vinosa* LAMARCK, 1822

Elongately biconic, spire height slightly less than $\frac{1}{2}$ total shell length, spire often shouldered. Sculpture variable, usually with broad rounded axial folds crossed by spiral ribs, the whole surface covered by fine, scaly, wavy, axial lamellae, but some specimens smooth or nodulose. Aperture weakly lirate, outer lip of mature specimens bears about 5 denticles on the inner margin. Columella smooth, angulate near the base of the short anterior canal. Exterior green-grey, or yellow, interior and columella brown-purple.

2 cm. Abundant. Intertidal rocks. N.S.W. to Albany, W.A. The species is extremely variable. The figured specimen is from Albany, W.A. The N.S.W. form was given subspecific status as *L. vinosa propinqua* TENISON WOODS, 1876. *L. reticulata* BLAINVILLE, 1832 is a similar species recognized by the presence of prominent shoulder nodules.

11. *Lepsiella flindersi* ADAMS & ANGAS, 1863

Spire high, whorls with angulate and weakly nodulose shoulders. Sculptured by irregular spiral ribs and fine longitudinal scale-like lamellae. Aperture with weak spiral lirae, which become denticles on the inner margin of the outer lip. Anterior canal wide, attenuate. Posterior canal notch lacking. Columella smooth. Exterior cream or greenish-white, aperture cream, white or brown, columella white or purple-brown.

3 cm. Common. Intertidal and shallow sub-littoral rocks. Vic. to Fremantle, W.A.

12. *Cronia avellana* REEVE. 1846

Spire high and sharply pointed, spire whorls nodulose. Body whorl sculptured by fine spiral striae, broad widely spaced longitudinal folds which become obsolete toward the lip, and a thick, scaled anterior fasciole. A narrow spiral groove encircles the body whorl anterior to the suture. Outer lip sharply edged and smooth, columella smooth, calloused posteriorly. Anterior canal notch deep, posterior canal shallow. Exterior white or cream with a black or dark brown spiral band at the sutures and another at the centre of the body whorl, anterior fasciole also black or brown, interior white or yellow, columella white.

3.5 cm. Common. Intertidal rocky shores. Cheyne Beach to Broome, W.A.

Plate 61. Thaidinae (descriptions of species 13 and 14 are on page 94) ($1\frac{4}{10}$ x natural size)

Muricidae, subfamily Thaidinae (Cont.)

Refer to page 93, plate 61,
(13, 14) for colour illustrations of shells.

13. *Cronia pseudamygdala* HEDLEY, 1903
Spire moderately high. Sculptured with broad, low axial folds, crossed by numerous fine, spiral ribs, with scaly axial lamellae in the interspaces. A thick, nodulose spiral fold encircling the whorls immediately anterior to the sutures, behind a shallow spiral furrow. Anterior fasciole low and short. Outer lip weakly dentate within, columella smooth. Light orange-brown with darker brown spiral bands on the body whorl, anterior fasciole dark brown, aperture and columella glossy cream or yellow.
   3.5 cm. Common. Intertidal rocky and coral reefs. Northern Qld. to central N.S.W.

14. *Drupella cornus* RÖDING, 1798
Ovate to biconic, spire high. Sculptured with spiral striae and spiral rows of high, rather pointed tubercles—4 rows on the body whorl, 2 on spire whorls. Anterior fasciole short but strong. Aperture smooth except for about 6 denticles on the inner margin of the outer lip. Anterior canal notch deep, posterior canal notch weak. Columella raised to form a shield, bears 3 or 4 weak anterior folds. Exterior, outer lip and columella white, deep interior white or yellow.
   4.5 cm. Common. Among rocks and corals, intertidal and in shallow water. Indo-West Pacific; Houtman Abrolhos, W.A. to eastern Qld. Cernohorsky (1969) reports that this species is sexually dimorphic, males being smaller than females, and have more bulbous shells ornamented with blunt nodules rather than pointed tubercles.

Thaidinae (Cont.) plate 62 (opposite page)

1. *Purpura persica* LINNAEUS, 1758
Spire moderately low, body whorl rounded, inflated and sculptured by numerous fine, rough cords. Anterior fasciole low. Outer lip finely denticulate, aperture large, oval, finely lirate. Anterior canal broad and shallow, posterior canal notch present. Columella smooth, broad and concave anteriorly. Exterior grey-brown with some of the spiral riblets appearing as dark brown lines interrupted by white spots. Inner edge of outer lip dark brown, deep interior white, lirations orange-brown. Columella yellow, white or orange.
   9 cm. Moderately common. Under stones and coral, intertidal and shallow sub-littoral. Indo-West Pacific; Qld. Records of this species from north W.A. need confirmation.

2. *Vexilla vexillum* GMELIN, 1791
Ovate to cylindrical, spire low. Smooth except for weak spiral striae. Outer lip finely denticulate. Anterior canal short, posterior canal notch deep. Columella weakly calloused and smooth. Exterior light brown or cream with wide, dark brown spiral bands, about 8 bands on the body whorl. Interior white, columella brownish.
   2.5 cm. Uncommon. Intertidal reefs. Indo-West Pacific; Barrow Is., W.A. to eastern Qld.

3. *Nassa francolina* BRUGUIÈRE, 1789
Ovate, spire moderately high. Whorls rounded, smooth except for numerous very fine spiral cords and a low rounded anterior fasciole. Outer lip finely denticulate, aperture smooth. Anterior canal broad, posterior canal not notched into lip but formed behind large opposing denticles on parietal wall and outer lip. Exterior red-brown with an irregular central spiral band of connected triangular whitish patches, with sometimes other less prominent bands anteriorly or posteriorly. Aperture cream, columella and inner margin of outer lip brown.
   7 cm. Common. Under stones and coral, intertidal and shallow sub-littoral. Indian Ocean; Houtman Abrolhos to Broome, W.A. This species closely resembles *N. serta*.

4. *Nassa serta* BRUGUIÈRE, 1789
Like *N. francolina* but distinguished by the presence of strongly beaded spiral ribs.
   7 cm. Common. Under stones and corals. Western Pacific including Qld., also at Cocos Keeling Is., but not recorded from W.A. This is the Pacific analogue of the Indian Ocean *N. francolina* with which it is often confused.

coral shells (FAMILY MAGILIDAE)

CORALLIOPHILIDAE AND HARPIDAE ARE ALTERNATIVE NAMES for this relatively small family closely related to the Muricidae. A great variety of shell forms occurs in the family probably because of the peculiar habits of the animals, and it is difficult to give diagnostic family shell characteristics. The animals live in close association with corals or other coelenterates. Some species live embedded inside massive corals, communicating with the outside only through small holes; others live in colonies attached to the outside of corals; some live attached to or embedded in anemones, soft corals or hydrozoans. These snails have no radula and probably feed by sucking out juices or mucous from the coelenterate host.

5, 5a. *Rapa rapa* LINNAEUS, 1767
Thin and bulbous, deeply umbilicate, spire almost flat, with a row of crowded rough irregular lamellae in the sutures. Body whorl sculptured by numerous spiral ribs which are stronger and further apart anteriorly where they bear low tubercular nodules. Anterior canal open, slightly projecting. Outer lip denticulate anteriorly. Columella smooth, reflected and partly covering the umbilicus anteriorly. White, sometimes faintly pink on the anterior canal.
   7 cm. Uncommon. Embedded at the base of soft corals. Indo-West Pacific; eastern Qld.

6, 6a. *Tolema australis* LASERON, 1955
Biconic, spire high, anterior canal long and straight. Whorls sculptured with close-set spiral rows of delicate hollow spines, broad and long on the angular shoulders. Outer lip spinose, outer wall of aperture with spiral channels, columella smooth. White.
   4 cm. Uncommon. Trawled off south-eastern Australia from N.S.W. to Vic.

7, 7a. *Coralliophila costularis* LAMARCK, 1816
Fusiform, with a high spire and attenuate anterior canal. Body whorl with 7 to 8 broad axial folds, crossed by numerous finely spinose spiral ribs. Anterior ribs coarser and more strongly spinose than posterior ribs. White or grey externally, deep lavender or purple in the aperture.
   6 cm. Uncommon. Lives on corals. Indian Ocean; Geraldton to Broome, W.A.

8. *Coralliophila violacea* KIENER, 1836
Ovate to bulbous, spire low, anterior canal narrow, curved, sometimes attenuate. Whorls rounded, sculptured by numerous fine, scaled spiral ribs, and a thick, raised, anterior fasciole. Aperture almost round with fine lirae on outer wall, outer lip finely denticulate. Columella smooth, reflected over a shallow umbilicus. White or grey externally, deep lavender or purple internally.
   4 cm. Common. Lives in colonies on massive corals. Indo-West Pacific; Geraldton, W.A. to eastern Qld.

9. *Quoyula madreporarum* SOWERBY, 1832
Thin, ovate, spire of medium height, low, or sometimes depressed. Whorls rounded, sculptured only by minute wavy spiral striae. Aperture wide and ovate to circular. Anterior canal and anterior fasciole wanting. Columella almost straight, smooth, with an angular margin. Exterior white, parts of the interior white, but purple at anterior end and on columella.
   3 cm. Common. Lives in colonies firmly attached to the prongs of *Pocillopora* and some other branching corals. Indo-West Pacific and also in the tropical central west coast of America; Pt. Cloates, W.A. to eastern Qld.

pagoda shells (FAMILY COLUMBARIIDAE)

ALTHOUGH THIS FAMILY has a long and rich fossil record, particularly in Australia and New Zealand, there are few living species. Only 3 living species are known from Australian waters, all of them from the outer part of the continental shelf or the continental slope off the eastern coast. The shells are fusiform, are generally prominently keeled, and have a long anterior canal, a paucispiral protoconch and a horny operculum. Little is known of the biology of the animals, but on shell and radula characteristics the family is placed close to the Muricidae. The family, including both fossil and living species, has been reviewed recently by Darragh (1969, Proc. Roy. Soc. Victoria, **83**: 63-119, pls 2-6).

Refer to page 95, plate 62 (10, 10a) for colour illustrations of shells.

10, 10a. *Columbarium spinicinctum* VON MARTENS, 1881
Spire high, whorls rounded but strongly keeled at the centre, keel with laterally directed spikes. Posterior slope of whorls and anterior slope of spire whorls smooth, anterior slope of body whorl with a low scaly spiral ridge or keel and a low spinose rib immediately posterior to it. Anterior canal long, slightly twisted, bears about 4 spiral ribs which may be spinose or scaly. Columella raised and smooth. Cream or pale brown with darker brown axial lines.

8 cm. Uncommon. Trawled in 70 to 125 fathoms off Cape Moreton, Qld. *C. caragarang* GARRARD, 1966 is a form of this species. *C. hedleyi* IREDALE, 1936 is another eastern Australian species which sometimes appears in collections. It may be distinguished from *C. spinicinctum* by its lower spire and the presence of 1 or 2 additional spinose spiral ribs between the anterior keel and the canal.

buccinid whelks (FAMILY BUCCINIDAE)

FEW GASTROPOD FAMILIES can match the Buccinidae in number of species. There are at least 2,000 species in this family including boreal (polar), temperate and tropical forms. Although there are many buccinids in Australia, the family has done better in other parts of the world. Many have rather drab shells, particularly the cold water species, but some small species are attractively coloured. All buccinids are carnivorous, some prey on other living molluscs, while many are scavengers and may be caught in traps baited with fish heads or meat. Female buccinids lay egg capsules, either singly or in masses. In some species larval development is direct, in others there is a planktonic larval stage.

1, 1a. *Penion maxima* TRYON, 1881
Fusiform, spire sharply acuminate, anterior canal open, very long, slightly upturned. Shoulders angulate, usually nodulose, whorls sculptured by many spiral cords crossed by longitudinal growth striae. Lip toothed, aperture lirate, columella smooth. Exterior fawn with a brown spiral band at the shoulders and sometimes lighter brown bands more anteriorly on the body whorl, aperture white.
25 cm. Moderately common. Trawled or taken in craypots off the east coast from southern Qld. to Tas. The Australian species have usually been placed in the genus *Austrosipho* but they appear to be congeneric with New Zealand species of *Penion*.

2. *Penion waitei* HEDLEY, 1903
Fusiform, spire acuminate, anterior canal long and open. Shoulders rounded, usually with elongate rib-like nodules, whorls sculptured with numerous strong spiral riblets. Lip weakly toothed, aperture lirate, columella smooth. Exterior and interior white.
16 cm. Uncommon. Trawled off south-eastern Australia from N.S.W. to Tas.

3. *Penion oligostira* TATE, 1891
Fusiform, whorls rounded, spire high and acuminate, anterior canal short. Early spire whorls nodulose, penultimate and body whorls sculptured only with crowded spiral cords. Lip profusely but finely toothed, aperture finely lirate, columella smooth. Exterior white, light brown or grey, spiral cords dark brown, aperture white.
10 cm. Moderately common. Western Vic. to S.A.

4. *Penion grandis* GRAY, 1839
Fusiform, anterior canal moderately long, open, slightly upturned. Shoulders angulate, usually nodulose, whorls sculptured by numerous spiral riblets. Outer lip toothed, aperture lirate, columella smooth. Exterior light brown with dark brown riblets, aperture white.
13 cm. Moderately common. Trawled or taken in craypots off southern Australia from Vic. to Tas. to the south coast of W.A.

5, 5a. *Phos senticosus* LINNAEUS, 1758
Ovate-oblong, with a high spire. Whorls shouldered, sculptured with strong axial folds crossed by spiral ribs and striae forming sharp nodules at the intersections. Aperture with strong spiral ribs, columella twisted, with 1 to 4 plaits or elongate nodules. Exterior white or light brown with brown spiral bands, interior and columella purple or mauve.
3.5 cm. Common in sand and on reefs. Indo-West Pacific; all States of Australia.

6, 6a. *Cantharus undosus* LINNAEUS, 1758
Biconic, outer lip thickened, anterior canal short. Whorls sculptured by angular spiral ribs and fine growth striae. Aperture spirally lirate, the lirations ending in denticles on the margin of the lip. The most posterior denticle large and located opposite an opposing parietal denticle. Columella with 3 or 4 thick folds. Exterior orange-brown with darker brown or black ribs interior and columella white. With a thick brown periostracum.
4 cm. Common. Under stones and corals in shallow water. Indo-West Pacific; Houtman Abrolhos, W.A. to eastern Qld.

7, 7a. *Cominella eburnea* REEVE, 1846
Biconic to fusiform, spire high and acuminate. Anterior canal a broad notch. Shoulders angulate with heavy nodules, often drawn out longitudinally to form broad ribs on the posterior part of the body whorl. Whorls also sculptured with low spiral cords. Aperture spirally lirate, columella smooth. Exterior light yellow-brown with irregular brown blotches, deep interior white, columella and inner margin of outer lip brown.
4 cm. Common. Intertidal. Vic. to Albany, W.A.

8, 8a, 8b, 8c. *Cominella lineolata* LAMARCK, 1809
Biconic to fusiform, spire high and acuminate, anterior canal a broad notch. Shoulders low, rounded and sometimes nodulose, whorls usually sculptured by low broad spiral riblets, but sometimes smooth. A broad shallow spiral groove present in front of the sutures. Aperture with 10 to 12 sharp spiral ribs on outer wall, columella smooth, calloused anteriorly. Colour variable, usually uniformly light yellow-brown, sometimes banded with dark brown and fawn spiral lines, or with interrupted brown lines giving a spotted appearance.
2.5 cm. Common. Intertidal. N.S.W. to southern W.A.

giant whelks and conchs (FAMILY MELONGENIDAE)

THIS IS A SMALL FAMILY of large shells, closely related to the Buccinidae. Galeodidae is an alternative name for the family. The best-known Australian representative is *Syrinx aruanus*, largest of all living gastropods. The crown conchs and busycon whelks of America are members of the same family. Most species are shallow water scavengers but some are said to take live oysters as prey. Females lay series of thick, plate-like egg-capsules held together by a gelatinous strand. In *Syrinx* the capsules are packed tightly together so that a concertina-like egg-mass is formed. Each capsule contains several hundreds of eggs but only 20 to 50 of these develop and hatch. The others serve as food or "nurse-eggs". Development is direct, i.e., the young hatch directly from the capsules as miniature snails. They have a very long, many-whorled protoconch but this is usually lost later in life (species are illustrated on plate 63) (opposite page).

Plate 63. Buccinidae and Melongenidae (descriptions of Melongenidae are on page 98) (⁹⁄₁₀ × natural size)

Melongenidae (Cont.)

Refer to page 97, plate 63,
for (9, 10) colour illustrations of shells.

9. *Volegalea wardiana* IREDALE, 1938
Elongately biconic to fusiform, anterior canal long, broad, open. Spire strongly shouldered, shoulders carinated and usually heavily nodulose. Whorls sculptured by spiral ribs. Aperture smooth, columella straight and smooth. Exterior brown, interior and columella rich orange-cream. Periostracum brown.

10 cm. Common on mud flats. Indo-West Pacific; Kimberley area in W.A. to northern Qld. The degree of nodulosity on the shoulders varies.

10. *Syrinx aruanus* LINNAEUS, 1758
Fusiform, anterior canal long, open and rather straight. Whorls sculptured by weak, rounded, longitudinal and spiral cords. Aperture and columella smooth. Umbilicus represented by an elongate slit. Exterior and interior rich creamy-yellow. Thick brown periostracum.

70 cm. Common. On intertidal flats and down to at least 20 fathoms. Northern Australia from Bunbury, W.A. to southern Qld. Northern shells have a high shouldered spire, and the shoulders are carinated sometimes nodulose (10); but at the southern end of the species range in W.A. the spire is low and straight-sided and the shoulders rounded. Formerly known under the generic name *Megalatractus* FISCHER.

band and spindle shells and their relatives

(FAMILY FASCIOLARIIDAE)

SOME RECENT AUTHORS RECOGNISE TWO FAMILIES, the Fasciolariidae which includes the band shells, latiruses and peristernias, and the Fusinidae (or Colidae) which includes the spindle shells. Little anatomical evidence supports this view and they are grouped in the one family here.

The family is widespread in both tropical and temperate seas. All species have a large horny operculum which fills the aperture when the animal withdraws into the shell, and the anterior canal is long, especially in the spindles. Members of the genera *Latirus* and *Peristernia* and similar forms have heavy shells with strongly ribbed and nodulose sculpture and several plaits at the anterior end of the columella. The band shells (*Pleuroploca*) are found on coral reefs. All these creatures are carnivorous and most feed on other molluscs. Spindle shells (*Fusinus* and related genera) have a smooth columella.

1. *Saginafusus pricei* SMITH, 1887
Fusiform, anterior canal moderately long. Whorls broad, rounded, sculptured by strong widely-spaced spiral ribs, the rib on the shoulder of the body whorl being nodulose. Spire whorls with low but strong longitudinal lamellae which connect the spiral ribs and give the surface a cancellate appearance. Aperture smooth, columella smooth and almost straight. Exterior light yellow-brown, interior cream or yellow.

15 cm. Moderately common. N.T. to northern N.S.W.

2. *Pleuroploca australasia* PERRY, 1811
Broadly fusiform, anterior canal moderately long. Whorls rounded with low rounded nodules on the shoulders and sculptured by numerous rough, rounded spiral cords. Aperture weakly lirate, columella smooth except for 2 or 3 plaits on the elbow at the base of the canal, parietal ridge well developed. Exterior light brown with darker brown spiral lines, aperture light brown. Covered by thin, shiny olive-brown periostracum.

15 cm. Common. N.S.W. to the south coast of W.A. The degree of angularity of the shoulders of the whorls is variable.

3, 3a. *Pleuroploca filamentosa* RÖDING, 1798
Fusiform, whorls slightly shouldered, anterior canal long and straight. Whorls rounded with low nodules on the shoulders and sculptured by numerous spiral ribs. Aperture with numerous spiral lirae, columella with 3 strong plaits at the base of the canal. Exterior dark brown with some ribs light brown or cream, and cream patches between nodules, interior orange-cream, columella orange, plaits cream.

11 cm. Common among corals. Indo-West Pacific; Qld.

4, 4a *Propefusus pyrulatus* REEVE, 1847
Fusiform, body whorl swollen and rounded, spire high, anterior canal long and curved. Whorls sculptured by low longitudinal folds, crossed by widely spaced spiral ridges with growth striae between them. Aperture weakly lirate, columella smooth or with the spiral ribs of earlier whorls showing through. Irregular parietal nodule borders the posterior siphonal notch. Exterior white, usually with pale brown spiral bands at the shoulders and sometimes anteriorly as well, or with pale brown patches on the shoulder nodules, aperture white.

8 cm. Common. N.S.W. to S.A.

5. *Fusinus novaehollandiae* REEVE, 1848
Elongately fusiform, spire high, anterior canal very long and straight. Whorls rounded, sculptured by weak, low, axial folds and sharp spiral riblets which are nodulose and most prominent on the shoulders. Columella calloused in adult specimens and bears transverse lirae. White or cream, usually covered by a brown fibrous periostracum.

30 cm. Moderately common in deep water. N.S.W. to S.A.

6. *Fusinus salisburyi* FULTON, 1930
Elongate, fusiform, anterior canal long and straight. Early spire whorls rounded with broad axial folds, reduced to angular nodules on the shoulders of the penultimate and body whorls. Strongly sculptured with angular spiral ribs and cords. Columella with a raised rim and bears transverse plaits. Umbilicus deep. Aperture wall strongly lirate, outer lip crenulate. White.

20 cm. Uncommon. Originally described from Japan but specimens (6) have recently been trawled at depths of 70-100 fathoms off Caloundra, Qld.

7, 7a. *Fusinus tessellatus* SOWERBY, 1880
Fusiform but less elongate than other fusinids, anterior canal comparatively short. Whorls rounded, spire whorls bear broad axial folds which usually become obsolete on the body whorl. Sculptured by about 10 to 12 roughly beaded spiral ribs with spiral cords between, and very fine axial cords. Aperture weakly lirate, columella smooth, bent at the base of the anterior canal. Ribs white or pale brown with regular dashes of dark brown, interspaces light brown with axially elongate streaks of white, interior white. Periostracum thick and brown.

6 cm. Uncommon. Among rocks. Southern W.A. from Esperance to Geraldton.

8, 8a. *Fusinus nicobaricus* LAMARCK, 1822
Fusiform, anterior canal straight and moderately long. Body whorl rounded with heavy nodules on the shoulders, several broad low rounded spiral ribs (also on the canal), and fine spiral and axial cords. Aperture weakly lirate, columella smooth or with weak plications. Exterior whitish with brown patches between the nodules and on the canal, aperture and columella white.

9 cm. Moderately common. Indo-West Pacific; N.T. to southern Qld.

9, 9a. *Fusinus colus* LINNAEUS, 1758
Elongately fusiform, spire high, anterior canal very long and straight. Body whorl rounded, earlier whorls with a central spiral row of large nodules. Whorls sculptured by many fine but strong spiral ribs. Aperture strongly lirate, columella callus smooth or so thin that ribs of earlier whorl are evident. White externally with brown patches between the nodules, particularly on the spire, tip of the anterior canal brown, interior white.

14 cm. Common. Indo-West Pacific; North-West Cape, W.A. to eastern Qld.

10. *Fusinus australis* QUOY & GAIMARD, 1833
Fusiform, spire high, body whorl broad; anterior canal long and straight. Whorls rounded, with slightly angulate shoulders, shoulders of the spire whorls with low nodules and axial folds but these are usually obsolete on the body whorl. Sculpture with angular spiral ribs and numerous weak axial cords in their interspaces. Columella smooth, straight except for a slight bend at the base of the anterior canal. Exterior brown; interior white. Periostracum thick and brown.

11 cm. Common. In shallow water, sand, among rocks and weed. Vic. to Geraldton, W.A.

Plate 64. Fasciolariidae ($\frac{8}{10}$ × natural size)

Fasciolariidae (Cont.)

1. *Latirus belcheri* REEVE, 1847
Broadly fusiform, anterior canal moderately long and straight. Spiral row of large flattened and ridged nodules encircles the shoulders, with another row of smaller nodules further anteriorly on the body whorl. Broad axial folds connect the posterior and anterior nodules. Whorls also sculptured with fine spiral ribs. Many weak lirae present on the outer wall of the aperture. Columella with a slight bend at the base of the anterior canal and several weak plaits at the bend. Exterior cream or white with irregular and interrupted axial bands of dark chocolate brown to black, interior white.
    7 cm. Common among rocks and corals. Indian Ocean; Port Hedland to Troughton Is., W.A.

2. *Latirus polygonus* GMELIN, 1791
Resembles *L. belcheri* but with nodules more rounded, spiral ribs stronger and more numerous. Exterior pale tan or flesh-brown with brown axial streaks.
    7 cm. Common among rocks and corals. Indo-West Pacific; eastern Qld.

3. *Latirus recurvirostris* SCHUBERT & WAGNER, 1829
Fusiform, spire high, anterior canal long with a thick anterior fasciole surrounding a deep umbilicus. Whorls bear heavy, spirally ridged nodules and spiral cords encircle the anterior canal. Outer lip toothed, aperture weakly lirate. Columella with a slight bend at the base of the anterior canal and weak plaits at the bend. Exterior fleshy-brown, with cream spiral ridges, axial rows of dark brown markings between the nodules, and a few interrupted spiral brown lines around the anterior canal. Interior and columella orange-brown.
    8 cm. Moderately common. Central Indo-West Pacific; Dampier Archipelago, W.A. to eastern Qld.

4. *Latirus pictus* REEVE, 1847
Ovately fusiform with a short anterior canal. Sculptured with low axial folds and weak spiral ribs, 1 rib on the shoulders and another around the centre of the body whorl are nodulose. Spiral striae present between the ribs. Aperture finely lirate, lip weakly toothed. Columella sharply bent at the base of the anterior canal with about 4 weak lirae at the bend. Exterior white with narrow yellow and brown spiral lines, interior glossy white, columella glossy white or brown-pink.
    5 cm. Uncommon. The original locality was Fiji. The specimen illustrated here is in the W.A. Museum collection labelled "Barrow Is." but the presence of the species in Australia needs confirmation.

5. *Latirulus turritus* GMELIN, 1791
Fusiform, slender, spire high, anterior canal short and narrow. Whorls rounded, with low rounded axial folds crossed by narrow spiral ribs. Outer lip toothed. Aperture finely lirate, columella sharply bent at the base of the anterior canal, with 3 or 4 plaits at the bend. Exterior rich brown, spiral ribs dark chocolate brown, interior and columella orange-brown.
    4 cm. Moderately common among rocks and corals. Indo-West Pacific; eastern Qld.

6. *Latirus paetelianus* KOBELT, 1876
Fusiform, moderately slender, deeply umbilicate, spire high, anterior canal of medium length. Whorls rounded, with prominent nodulose axial folds crossed by numerous spiral cords. Outer lip finely toothed, aperture lirate, columella with a band at the base of the anterior canal and 2 or 3 weak plaits at the bend. Exterior, inner margin of lip, and columella uniform brown (ochraceous), deep interior white.
    6 cm. Common among rocks. Central Indo-West Pacific; Dampier Archipelago, W.A. to the Gulf of Carpentaria. Hedley (1912) described the Australian form as a distinct subspecies which he named *carpentariensis*.

7. *Latirus gibbulus* GMELIN, 1791
Fusiform, spire high, anterior canal broad but moderately short. Heavy, rounded nodules on the shoulders at the whorls become low axial folds further anteriorly. A broad, low, flattened anterior fasciole encloses a wide and deep umbilicus. Outer lip toothed, aperture finely lirate, columella smooth, almost straight. Exterior rich red-brown with darker spiral lines, interior cream, columella brown-cream.
    7 cm. Moderately common. Indo-West Pacific; eastern Qld.

8. *Latirus walkeri* MELVILL, 1895
Fusiform, slender, usually slightly umbilicate, spire high, anterior canal of medium length and straight. Whorls with prominent broad axial folds crossed by spiral cords. Outer lip weakly toothed, aperture weakly lirate, columella with only a slight bend at the base of the anterior canal and several very weak plaits at the bend. Axial folds yellow or white, interspaces light or dark brown, tip of anterior canal light or dark brown, columella white or faintly purplish, interior white.
    3 cm. Moderately common among rocks. North-West Cape to Broome, W.A.

9. *Peristernia nassatula* LAMARCK, 1822
Biconic, anterior canal moderately short. Whorls with about 10 broad axial folds crossed by numerous spiral ribs giving the surface a rough nodulose texture. Strong lirae on the outer wall of the aperture. Columella sinuous with 3 plaits at the bend near the base of the anterior canal. Exterior brown, axial folds white, spire apex rose pink, interior columella and tip of anterior canal mauve.
    3.5 cm. Common. Indo-West Pacific; N.T. and eastern Qld.

10. *Peristernia incarnata* DESHAYES, 1830
Elongately biconic, anterior canal projecting but short. Whorls with about 10 smooth, thick, rounded axial ribs with numerous radial lamellae between. Aperture weakly lirate, columella smooth, sharply cornered at the base of the anterior canal. Axial ribs rich yellow-brown, interspaces darker brown, aperture pinkish or mauve.
    3 cm. Common. Northern Australia, Geraldton, W.A. to eastern Qld.

11, 11a. *Peristernia australiensis* REEVE, 1847
Biconic, anterior canal projecting but short. Whorls sculptured with about 8 broad rounded axial ribs and numerous sharp spiral cords which cross the ribs and the interspaces. Aperture with strong spiral ribs on the outer wall, columella with 2 or more transverse ridges on the angular corner at the base of the anterior canal. A prominent parietal nodule borders the posterior siphonal notch. Exterior whitish with brown blotches between ribs, aperture brown to mauve.
    3.5 cm. Common. Indo-West Pacific; N.T. to eastern Qld.

12. *Latirolagena smaragdula* LINNAEUS, 1758
Broadly ovate, anterior canal short. Whorls rounded, smooth except for very fine and low spiral cords. Aperture finely lirate, columella sinous with 2 to 3 weak plaits at the base of the canal. Exterior brown with fine darker brown spiral lines, inner margin of outer lip brown, interior and columella glossy blue-white.
    4 cm. Common among rocks and corals. Indo-West Pacific; eastern Qld.

Plate 65. Fasciolariidae (1$\frac{3}{10}$× natural size)

dog whelks (FAMILY NASSARIIDAE)

THE NASSARIIDAE IS A LARGE FAMILY with many species in temperate and tropical seas although most are tropical. Most species are very variable and the present taxonomy of the family, at both specific and generic level, is quite unsatisfactory. A selection of only a dozen species has been illustrated here but many more species than this inhabit Australian waters. Dog whelks are active creatures with a large foot bearing 2 "tails" at its posterior end. Many species are abundant on marine or estuarine intertidal mud or sand flats, others live on mud or sand bottom in deeper water. These scavengers may be kept alive in aquaria or dishes of sea water for long periods by feeding on cracked bivalves. The females lay small egg-capsules which they attach to the substrate or other shells. There is a long-lived planktonic larval stage. The shell of a dog whelk is usually ovate, often strongly sculptured, without an umbilicus, and usually with a spiral groove or fossula behind the short slot-like anterior canal. The operculum is small and horny.

1, 1a. *Nassarius arcularius* LINNAEUS, 1758
Ovate, spire of medium height. Parietal wall and columella so heavily calloused that the shell has a wide thick flat base. Spire whorls with heavy axial ribs, on the body whorl the axial ribs reduced and rather sharp edged. Heavy tubercles present on shoulders of the body whorl. Aperture lirate, columella curved, smooth except for several anterior nodules. Parietal nodule prominent. Exterior white or cream, often with brown spots between the shoulder tubercles, interior brown or yellow.
   3.5 cm. Common. Indo-West Pacific; eastern Qld.

2, 2a. *Alectrion papillosus* LINNAEUS, 1758
Elongate, spire high, whorls sculptured with spiral rows of large rounded nodules. 5 or 6 radiating spikes present on the anterior part of outer lip. Aperture minutely spirally striate, columella smooth, weakly calloused. Exterior light yellow-brown or cream with patches of brown, nodules white, spire apex pink, interior white.
   5 cm. Moderately common. Indo-West Pacific; eastern Qld.

3, 3a. *Nassarius coronatus* BRUGUIÈRE, 1789
Ovate, spire high. Early apical whorls and the body whorl near the margin bear axial ribs, remainder smooth and glossy. Shoulders of the whorls nodulose. Aperture lirate, columella and underside of body whorl heavily calloused, columella with some weak nodules or ridges anteriorly. Parietal nodule prominent. Exterior cream or pale yellow-brown, sometimes with a brown central spiral band, interior brown, ventral callus white.
   3 cm. Common. Indo-West Pacific; Dampier Archipelago, W.A. to eastern Qld.

4, 4a. *Niotha pyrrhus* MENKE, 1843
Ovate, spire high. Sculptured by radial and spiral cords with prominent granules at the points of intersection. Outer lip slightly thickened, usually denticulate within. Aperture lirate, columellar callus weak and granulose. Exterior white with a brown spiral band at the anterior end, and another near the sutures on the spire whorls which becomes a central spiral band on the body whorl. The central band may be very wide or divided into 2 or more bands. Columella and outer margin of aperture brown.
   2 cm. Common. Vic. to Fremantle, W.A.

5, 5a. *Parcanassa pauperata* LAMARCK, 1822
Ovate, spire of medium height. Whorls with strong axial ribs crossed by weaker spiral cords giving a granulose appearance, spiral cords strongest anteriorly. Outer lip thickened and strongly toothed on its interior margin. Columella curved, heavily calloused, weakly nodulose anteriorly. Exterior cream or brown, usually with a darker brown broad central spiral band on the body whorl. Columellar callus glossy, yellow or white, anterior end purple-brown.
   2.5 cm. Common. Lives in estuaries and protected bays. Vic. to Geraldton, W.A.

6, 6a. *Plicarcularia granifera* KIENER, 1834
Ovate, spire of medium height. Parietal wall and columella so heavily calloused the shell has a wide, thick and flat base. Ornamented with spiral rows of heavy tubercles, 2 rows visible on the spire whorls, 4 on body whorl. Outer lip thickened. Aperture lirate, columella smooth, but tubercles on the parietal wall showing through the overlaying callus. Exterior ground colour dull cream, tubercles white, callus glossy white.
   2 cm. Common. Indo-West Pacific; eastern Qld.

7, 7a. *Niotha bicolor* HOMBRON & JACQUINOT, 1853
Ovate, spire moderately high with impressed sutures. Sculpture of fine axial and spiral ribbing forming a rough granulated surface. Aperture and inner margin of the columella strongly lirate. Thick columellar callus present over part of the parietal wall. Spire apex blue-grey, becoming whitish on later spire whorls and body whorl, aperture and callus glossy white.
   2 cm. Common. Indo-West Pacific; Pt. Cloates, W.A. to N.S.W.

8, 8a, 8b. *Zeuxis dorsatus* RÖDING, 1798
Ovate, spire high and conical. Whorls rounded, early apical whorls with axial ribs, later spire whorls and body whorl smooth and glossy, with deep spiral striae at the anterior end. Outer lip thickened, and bears 6 to 8 small teeth anteriorly. Columella weakly calloused, smooth except for 1 transverse rib posteriorly. Aperture lirate. External colour variable, may be uniform blue-grey, green-grey or brown (sometimes almost black), weak banding sometimes evident. Interior brown, columella white.
   3.5 cm. Common. Indo-West Pacific; Exmouth Gulf, W.A. to eastern Qld.

9, 9a, 9b. *Plicarcularia thersites* BRUGUIÈRE, 1789
Very solid, ovate, spire rather short. Heavily calloused columella and outer lip form a broad flat base, dorsal side peculiarly humped. Whorls sculptured with strong axial ribs, interrupted by beaded spiral cords anteriorly. Inner margin of outer lip toothed, columella also usually toothed. Exterior grey or brown with a central yellow band on the body whorl, basal callus glossy and white or yellow.
   2.5 cm. Common. Indo-West Pacific; eastern Qld.

10, 10a. *Niotha gemmulata* LAMARCK, 1822
Ovate to globose. Spire moderately high, sutures deeply impressed. With a heavy sculpture of rounded nodules regularly arranged in axial and spiral rows, the nodules of the first spiral row in front of the sutures larger than the others. Columellar callus weak. Exterior light yellow-brown, often with darker brown patches.
   3.7 cm. Moderately common. Indo-West Pacific; Broome, W.A. to N.S.W.

11, 11a. *Alectrion glans* LINNAEUS, 1758
Elongate, spire high, sutures impressed. Early spire whorls granulated or axially ribbed, shoulders of whorls weakly nodulose, shell otherwise smooth and glossy. Outer lip thin, bearing 4 or 5 short spikes anteriorly. Aperture weakly lirate, columellar callus weak, columella smooth except for a prominent parietal nodule. Exterior cream, pale yellow-brown or blue-grey, suffused with orange-brown, and with a number of distinct fine red spiral lines. Interior and columella cream.
   5 cm. Common. Indo-West Pacific; Geraldton, W.A. to eastern Qld.

12, 12a. *Alectrion rufula* KIENER, 1834
Ovate, spire of medium height. Whorls rounded, sometimes slightly shouldered at the sutures, smooth except for weak axial ribs on the apical whorls and weak spiral striae anteriorly. Outer lip sharp with a few weak teeth anteriorly. Outer wall of aperture smooth, columella weakly calloused, smooth. External colour variable, ground colour white, cream or pale orange, overlain by irregular axially-aligned blotches of brown or orange, and with thin orange or brown spiral lines. Spire apex pink or purple, columella white.
   2.5 cm. Common. W.A. from Albany to Geraldton. This species lacks the shoulder nodules of its northern relative *A. glans*, and is more globose than the eastern species *A. particeps* HEDLEY, 1915, but the relationships of these species need study.

Plate 66. Nassariidae (1³⁄₁₀ˣ natural size)

Plate 67. *Oliva miniacea* Röding, 1798. At a depth of 6 feet, off Mossman, N. Qld.   (*Photo*—Neville Coleman).

olives and ancillids

(FAMILY OLIVIDAE)

THE FAMILY OLIVIDAE contains about 60 living species and many subspecies and distinctive forms which live in tropical or warm seas. Four subfamilies have been recognized and of these, 2 are well represented in Australia, i.e. the Olivinae or true olives, and the Ancillinae or ancillas. However, the anatomy of many Australian species remains unknown and some may eventually be placed in either of the other 2 families, the Olivellinae and the Agaroniinae which are best represented in the Americas. True olives are most numerous in the warm waters of northern Australia but there is 1 species on the southern coast. The ancillas, on the other hand, are numerous in the south as well as in the north.

Most true olives (genus *Oliva*) are extremely variable in shell colour and pattern; several different species have dark chocolate-brown colour forms. For this reason collectors usually have great difficulty in identifying the species, but the recent publication by Zeigler and Porreca (1969) should be a great help (see references). The extremely glossy surface of olive shells rivals that of cowries. The shells are solid and, besides their gloss, are characterized by their cylindrical shape, low spire and long narrow aperture with several columellar plaits. The head and foot are large and muscular for burrowing. There is no operculum.

Ancillids have a thinner, less glossy shell which usually has a wider aperture and higher spire, and there may be a small operculum. Often the sutures are covered with a callus.

Useful characters used in the identification of olives and ancillas concern the sculpturing of the underside. There is usually an anterior zone on the body whorl demarcated by an incised groove or raised ridge. This zone is called the "fasciolar band" and may be wide or narrow, smooth or

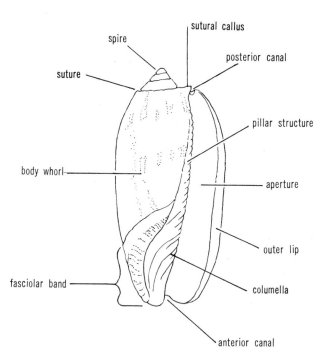

Fig. 19. *Oliva* sp. showing major external parts of shell.

grooved. The parietal wall and columella are usually calloused and the callus is called the "pillar structure". It may be smooth or transversely lirate.

Members of the family Olividae are active snails which crawl about, sometimes on top of the sand but more commonly below the surface with only the tip of the siphon showing. There have been reports that some species are capable of swimming, probably as an escape reaction to avoid predators. Wilson (1969) described such behaviour in *Ancillista cingulata* SOWERBY, 1830. In this species the large forepart of the foot (propodium) was the propelling organ. The propodium repeatedly flapped up and down, first dorsally and then ventrally, at regular intervals of slightly more than one second. This action, maintained for about 45 seconds, propelled the animal to the surface of the water in a large dish and around in an erratic manner. Swimming occurred spontaneously or as a result of agitation. Recently there have been reports that *Amalda elongata* swims in a similar manner (Mrs Fay Back, personal communication). Some other species are said to swim by flapping the side flaps of the foot. All the members of this family are scavengers or predators and as they crawl about under the surface of the sand searching for food they leave behind a characteristic trail. "Tracking" for olives is a very successful way of collecting them. Another successful method is to pin fish heads or other meats to the sea bed and revisit this bait at intervals. Olives can "smell" such bait from several yards away and will gather in numbers in the sand around and beneath it. There have also been reports that some of the larger species have been caught on a baited hook and line.

A recent publication by Olsson and Crovo (1968) has described the feeding habits of the American species *Oliva sayana* in aquaria as follows: "If pieces of fish, shrimp, or steak be placed in the tank, the Olives soon become aware of them and will begin to emerge and start on a round of investigation. When the food morsel is discovered, it is quickly seized and pushed back under the foot and infolded in its hinder section as if tucked away in a pocket, which

swells up into a rounded ball, the food item hidden away so completely that no part of it is visible externally. The Olive then retreats below sand level, going down head first. The same procedure is used with live mollusks such as *Donax* and *Laevicardium,* to which latter *Oliva sayana* is especially partial."

Little is known about the breeding habits of members of this family. Information is available on egg-laying and development of *O. sayana,* again through aquarium observations reported by Olsson and Crovo (1968). This species lays small spherical capsules which are shed individually into the water. Each capsule holds from 20 to 50 embryos which hatch after about 7 days as free-swimming veligers. Of special interest is the fact that the egg capsules are not attached to the substrate but lie free on the surface of the sand. They are probably subject to dispersal by water currents at this stage. Dr Olsson and Mrs Crovo have remarked that olives are very adaptable animals which are easy to keep in aquaria. The kind of information which they have obtained from their aquarium studies is of great interest to naturalist and scientist alike. Perhaps some Australian collectors could make similar studies of Australian olives?

Selected references:

BURCH, JOHN Q. & BURCH, ROSE L. (1967). The family Olividae. Pacific Science, **21** : 503-522.

OLSSON, AXEL A. & CROVO, L. E. (1968). Observations on aquarium specimens of *Oliva sayana* RAVENEL. Veliger, **11** : 31-32, pl 2.

WILSON, BARRY R. (1969). Use of the propodium as a swimming organ in an ancillid (Gastropoda : Olividae). Veliger, **11** : 340-342, pl 53.

ZEIGLER, ROWLAND F. & PORRECA, HUMBERT C. (1969). *Olive Shells of the World.* W. Henrietta, N.Y.

Plate 68. *Oliva* sp. Pt. Cloates, W.A. (see species 1, 1a, page 106). *(Photo—*Barry Wilson).

1, 1a. *Oliva* sp.

(see also colour plate of live animal, page 105, plate 68)

Very like *O. lignaria* but tending to be broader, less cylindrical and more solid. Penultimate whorl thickly calloused (in adult specimens), aperture white or only faintly blue-tinted. Most with a ground colour of gold or yellow overlain by thick clouded brown zigzag axial lines, sometimes darker and more distinct brown markings present in 3 spiral bands. Some specimens have a white or greenish ground colour.

7 cm. Common. Wedge Is. to North-West Cape, W.A. Living specimens of this species may be found on the same sand flats as *O. lignaria* in the Pt. Cloates area but we believe them to be a distinct species. They may be the problematical *O. irasans* LAMARCK, 1822.

2, 2a, 2b. *Oliva lignaria* MARRAT, 1868

Slender, rather cylindrical. Spire of medium height, sutures grooved. Fasciolar band short, usually poorly defined. Pillar structure thick, reaches about ⅔ of the way to the posterior end of the aperture. Lirae confined to the pillar structure. Anterior end of the pillar structure crossed by 3 or 4 sharp-edged revolving ridges. External colour variable, usually white with brown zigzag axial lines and 3 darker spiral bands. Fasciolar band with separated crescent-shaped brown lines, pillar structure white, aperture bluish or purple-grey. Base colour sometimes cream, and the zigzag lines dark brown and dense. Entirely chocolate brown specimens not uncommon.

5 cm. Common. Indo-West Pacific; Geraldton to Broome, W.A. This species is popularly known as *O. ornata* MARRAT, 1867 but according to Burch & Burch (1967) the name is preoccupied and *O. lignaria* is the name that should be used.

3, 3a, 3b. *Oliva annulata* GMELIN, 1791

Broad, spire moderately high, sutures deeply channelled. Fasciolar band raised and demarcated by an incised spiral line. Pillar structure heavily striate reaching almost to the posterior end of the aperture, 2 sharp-edged revolving ridges arising from the pillar structure present on the anterior end of the fasciolar band. Exterior usually cream or fawn-yellow with a fawn band in front of the sutures, a spiral row of widely separated blue-brown spots at the sutures, numerous diffuse rather triangular spots on the body whorl and a spiral row of large crescent-shaped blue-brown spots on the fasciolar band. Rarely uniformly creamy-yellow (3b). Interior yellow-orange.

6 cm. Moderately common. Indo-West Pacific; eastern Qld.

4, 4a, 4b, 4c, 4d, 4e. *Oliva carneola* GMELIN, 1791

Stout, spire low, sutures deeply channelled. Fasciolar band demarcated by an incised line, pillar structure and its lirae do not reach the posterior end of aperture. Anteriorly 3 or 4 sharp edged revolving ridges present. Colour variable, usually white with 1 broad posterior, 1 broad anterior and 1 narrow central spiral band of orange or yellow or more rarely brown, bands sometimes irregular, sometimes absent. Fasciolar band, pillar structure and interior white.

2 cm. Common. Indo-West Pacific; eastern Qld.

5, 5a. *Oliva caldania* DUCLOS, 1835

Small, stout, sutures grooved. Fasciolar band a raised ledge with a sharp posterior edge. Pillar structure reaches only half way back to posterior end of aperture. Lirae confined to the pillar structure consisting of 3 or 4 strong lirae posteriorly, and 3 very strong ridges with deep grooves between them running around the anterior end. Exterior creamy white, yellow or yellow-brown with longitudinal zigzag brown lines. Fasciolar band the same ground colour with wavy transverse brown lines. Pillar structure white, interior blue.

2 cm. Common, Central Indo-West Pacific; North-West Cape, W.A. to eastern Qld. Also known as *O. brettinghami* BRIDGMAN, 1909.

6, 6a. *Oliva caerulea* RÖDING, 1798

Thick, broad, sutures channelled. Fasciolar band demarcated by a raised spiral cord. Pillar structure reaches only about ½ way along the aperture, but short teeth or lirae present on posterior end of the parietal wall. Four broad ridges, sharply angular at their posterior edges, rise from the pillar structure and run spirally around the anterior part of the fasciolar band. Exterior white or yellow, with black streaks rising from sutures and diffuse V-shaped zigzag greenish or blackish axial lines or spots edged with yellow and concentrated in 3 spiral bands. Fasciolar band with a spiral row of thin crescent-shaped marks, pillar structure cream, orange or white. Interior dark violet.

6 cm. Common. Indo-West Pacific; North-West Cape, W.A. to eastern Qld. The species is widely known as *O. episcopalis* LAMARCK, 1810.

7, 7a. *Oliva reticulata* RÖDING, 1798

Solid and broad. Spire low, suture channelled. Thick parietal callus present beside the posterior canal. Fasciolar band poorly defined by a thin sharp incised line. Pillar structure thickly calloused, reaching about ⅔ of the way to posterior end of the aperture. Lirae confined to the pillar structure, sometimes obsolete even there. 3 or 4 sharp-edged revolving ridges present at the anterior end of the pillar structure. Exterior cream or greenish with crowded zigzag brown lines and 2 more dense bands posteriorly. Posterior part of the fasciolar band with distinct transverse brown lines. Aperture white, pillar structure dark orange or orange-red.

4.5 cm. Common. Indo-West Pacific; Qld. Also known as *O. sanguinolenta* LAMARCK, 1811, and *O. variegata* RÖDING, 1798. The illustrated specimens are from the Philippines. *O. vidua* RÖDING, 1798 is a similar shell common in Qld. It is very squat, with a thick parietal callus but the spire is very low or almost flat, the reticulate pattern is lacking or less dense, and the columella is white or pale red.

8, 8a *Oliva miniacea* RÖDING, 1798

(see also colour plate of live animal, page 104, plate 67).

Sutures deeply channelled. Fasciolar band a raised shelf with a sharp posterior edge. Pillar structure reaches at least ⅔ of the way to the posterior end of the aperture, lirae present along whole length of the aperture. Anteriorly the pillar structure thick, with 3 high, angular revolving ridges. Body whorl creamy-white with wavy axial lines of orange and blue-green or blue-grey, and 2 purplish spiral bands. Spire whorls and the posterior part of the fasciolar band with crowded purplish lines. Pillar structure cream, interior deep orange.

8 cm. Common. Indo-West Pacific; Qld. A large, beautiful and well known species often called *O. erythrostoma* MEUSCHEN, 1787 or *O. sericea* RÖDING, 1798.

9, 9a. *Oliva tessellata* LAMARCK, 1811

Stout, spire whorls calloused, sutures deeply channelled. Fasciolar band demarcated by a sharp, thin incised line. Pillar structure reaches almost to the posterior end of the aperture and bears thick transverse folds and 2 or 3 sharp-edged revolving ridges on the anterior end. Exterior cream, sometimes green-cream, with widely spaced purple-brown spots. Fasciolar band unmarked, outer lip and anterior end of the pillar structure white, aperture and remainder of pillar structure deep violet.

3 cm. Uncommon. Indo-West Pacific; Qld.

10, 10a, 10b. *Oliva oliva* LINNAEUS, 1758

Spire moderately high, sutures deeply grooved. Fasciolar band short, poorly defined, pillar structure thickly calloused anteriorly and reaches slightly more than ½ way along the aperture. Lirae confined to the pillar structure which at its anterior end is crossed by 3 strong sharp-edged revolving ridges. Exterior cream or brown with brown or purplish spots, zigzag lines and sometimes spiral bands. Fasciolar band sometimes with indistinct crescent-shaped lines, pillar structure white, aperture pale brown or blue.

4 cm. Common. Indo-West Pacific; Wedge Is., W.A. to southern Qld. The popular name *O. ispidula* LINNAEUS, 1758 is misapplied to this species.

Plate 69. Olividae (1²⁄₁₀ × natural size)

Olividae (Cont.)

1, 1a, 1b. *Oliva australis* Duclos, 1835
Elongate, spire high, sutures channelled. Fasciolar band demarcated by an incised line, pillar structure does not extend as far as the posterior end of aperture. Lirae confined to the pillar structure, with 4 strong sharp-edged revolving ridges at its anterior end. 2 colour forms exist: (i) white with interrupted zigzag brown axial lines or diffuse spots, bluish or brown flames at sutures, a brown fasciolar band, interior white (1b); (ii) darker, ground colour is green-yellow with crowded zigzag lines or triangular marks but rarely spots, bluish flames at the sutures, a dark brown fasciolar band, interior blue-grey (1, 1a). Pillar structure white in both forms.
  3 cm. Common. Vic. and Tas. to Broome in the north of W.A. The relationship of the 2 colour forms needs investigation.

2, 2a, 2b. *Oliva mustellina* Lamarck, 1811
Slender, cylindrical, spire low, sutures deeply channelled. Parietal wall beside the posterior canal calloused. Fasciolar band demarcated by an incised line. Pillar structure reaches almost to the posterior end of the aperture, with strong transverse folds throughout its length and 3 strong sharp-edged revolving ridges at the anterior end. Exterior cream, grey or light brown with crowded irregular zigzag lines and small spots at the sutures. Fasciolar band with crowded crescent-shaped brown lines. Pillar structure white, sometimes orange anteriorly. Interior blue-white or violet.
  4 cm. Common. Indo-West Pacific; Qld.

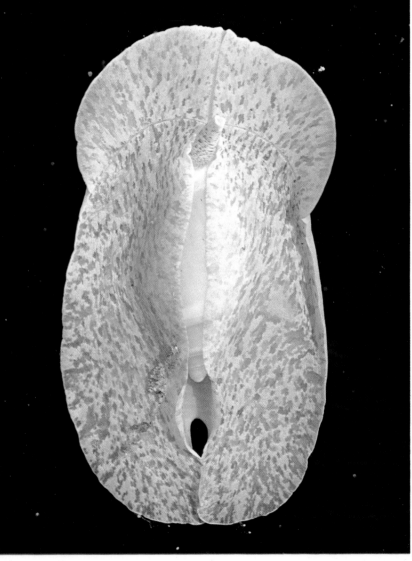

Plate 70. *Ancillista cingulata* Sowerby, 1830, Pt. Cloates, W.A.
(*Photo*—Barry Wilson).

3, 3a. *Ancillista cingulata* Sowerby, 1830
Thin, ovately fusiform, spire high with rounded apex. Whorls glossy and smooth except for minute spiral striae, a fine shallow incised spiral pre-sutural groove and a narrow raised spiral ridge on the spire whorls behind the sutures. Fasciolar band narrow, demarcated by a fine incised spiral groove and bears a single low spiral ridge. Pillar structure lacking. Whorls creamy grey or fawn, spire whorls sometimes blue-grey. Zone between the sutures and the pre-sutural groove white. Narrow brown spiral band present behind the fasciolar groove and on the posterior part of the fasciolar band. Interior brown.
  6 cm. Moderately common. Northern Australia; Shark Bay, W.A. to north-eastern Qld.

4. *Ancillista velesiana* Iredale, 1936
Very like *A. cingulata* but more swollen, with the spire whorls covered by a thin glossy brown callus. Pre-sutural zone also brown but a narrow white spiral band present at the pre-sutural groove.
  9 cm. Moderately common. Taken by prawn trawlers off southern Qld. and N.S.W.

5. *Alocospira marginata* Lamarck, 1811
Moderately thin, ovately fusiform, spire high. Body whorl smooth except for minute spiral striae and an incised spiral pre-sutural groove. Spire whorls with strong spiral ridges. Fasciolar band demarcated by an incised spiral line ending in a pointed tooth on the outer lip, and with two broad sharp-edged revolving ridges. Pillar structure narrow anteriorly but broad posteriorly where it usually extends over and beyond the sutures. Columella twisted, with 6 or 7 fine spiral cords. Exterior cream, yellow or orange, with brown patches in the spiral zone between the sutures and the pre-sutural groove.
  4 cm. Moderately common. N.S.W. to S.A. including Tas.

6, 6a. *Alocospira monilifera* Reeve, 1864
Moderately thin, ovately fusiform, spire high. Body whorl smooth except for minute spiral striae and an incised spiral pre-sutural groove. Fasciolar band demarcated by a spiral groove, and has 2 broad raised sharp-edged revolving ridges. Pillar structure indistinct, columella twisted and has 3 or 4 spiral ridges. Body whorl white with fine zigzag light brown lines and strong darker brown streaks in the pre-sutural zone. Transverse brown bars present on the anterior part of the fasciolar band but the posterior part white or yellow. Spire sometimes stained yellow (6).
  2.5 cm. Common. Vic. to Fremantle, W.A. There are several small ancillid species like this in southern Australia.

7, 7a. *"Amalda elongata* Gray," 1847
Thin, elongate, cylindrical. Aperture length about ¾ of the total, anterior canal broad. Whorls with fine spiral striae except on the smooth pre-sutural zone and fasciolar band. Pre-sutural groove a fine incised spiral line. One weak spiral rib on fasciolar band, demarcated by a fine incised spiral line. Pillar structure indistinct. White, sometimes with a brown-stained spire apex.
  5 cm. Common. Northern Australia; Shark Bay, W.A. to northern Qld. (?). Although most recent authors have placed this species in the genus *Amalda* it probably does not belong there.

8, 8a. *Zemira australis* Sowerby, 1841
Rather solid, ovate, with a high strongly shouldered spire. Sutures deeply and widely channelled. Whorls sculptured with numerous fine spiral striae. Spiral groove present on body whorl slightly in front of centre and ending in a short tooth on the outer lip. Columella calloused but smooth. Cream with yellow and brown blotches and streaks, particularly at the sutures. Aperture and columella white.
  3 cm. Uncommon. N.S.W. to Vic. and Tas. Some authors exclude this genus from the Olividae and place it in a family of its own. *Z. bodalla* Garrard, 1966 is another species from southern Qld. distinguished by its coarser sculpture. Do not confuse this shell with the cancellarid *Nevia spirata* Lamarck, 1843 which is remarkably similar but lacks the spiral groove on the body whorl and has strong lirae on the aperture and strong columellar plaits.

Plate 71. Olividae (1⁹⁄₁₀ × natural size) [

harp shells (FAMILY HARPIDAE)

THIS SMALL FAMILY includes some of the most attractive shells in the ocean. They are usually shiny, beautifully coloured and patterned and sculptured with strong axial ribs more highly polished than the remainder of the shell. The anterior siphon canal is a deep, wide notch, the columella is without folds and there is no operculum. The foot is broad and the front end of it greatly expanded laterally to form a wide, flat and muscular organ which aids in burrowing. There is a long respiratory siphon and, although the head is small, long eye stalks.

Harps are active, carnivorous molluscs which burrow in sand in depths ranging from low tide level to many fathoms. Little is known of their biology. At least some species are known to practice "autotomy" when being attacked, breaking off the rear part of the foot to make off, leaving that writhing part behind to distract the attacker. The family contains only about a dozen living species, most of which live in the Indian and Pacific Oceans. Only *Harpa amouretta*, *H. harpa* and *H. articularis* have so far been recorded on Australia's northern shores but other Indo-West Pacific species may yet be found there.

In deep water in southern Australia are 2 small fragile and very delicately sculptured species named *exquisita* and *punctata*, which are seldom seen. A third species of the same group has been dredged recently off Fremantle in Western Australia and at present awaits description. The generic classification of these small harps has been a subject of some disagreement among Australian taxonomists, but the name *Austroharpa* is now generally used for them. They are surviving species of a group of harps which flourished in southern Australia in middle Tertiary times. In South Australia and Victoria there are about 10 fossil species referable to *Austroharpa*, and one other which belongs to the related extinct genus *Eocithara*.

True *Harpa* are confined to tropical waters and have a tall conical protoconch consisting of several whorls. The protoconch of the living and fossil *Austroharpa* is dome-shaped and consists of little more than 1 whorl. In some of the fossil species the protoconch is bulbous. So far nothing is known of the anatomy of *Austroharpa* but when living specimens are obtained for study they will provide much interesting data for comparing these ancient cool water harps with their more modern cousins of tropical waters.

There is no recent revision of the Harpidae. A summary of the taxonomy of the recent and fossil *Austroharpa*, with references to the earlier literature, is given by Cotton and Woods (1933), Rec. S.A. Museum, 5 (1): 45-47.

Fig. 20. *Harpa amouretta* RÖDING, 1798, Qld., drawn from colour photograph of a living animal.

1, 1a. *Harpa articularis* LAMARCK, 1822

Rather thin and inflated. Width equal to ⅔ the length. Spire of medium height. Each whorl with about 15 axial costae. Exterior fawn with undulating white and brown lines between the ribs, ribs darker fawn or pale orange crossed by a few rather thick brown lines. Parietal and columellar callus with a large but undivided dark brown blotch.

10 cm. Moderately common. Indo-West Pacific; Onslow, W.A. to eastern Qld. Shells from Qld. have been misidentified as *H. davidus* RÖDING, 1798, a different species found in the Indian Ocean. *H. articularis* resembles *H. major* RÖDING, 1798 (2) which may be distinguished by its heavier shell, a divided parietal and columellar blotch, and red colour.

2. *Harpa major* RÖDING, 1798

Not yet recorded from Australian waters, but expected to occur in the far north.

10 cm. Uncommon. Indo-West Pacific. The specimen illustrated is from Mauritius and is included here for comparison with *H. articularis* (see remarks under that species).

3, 3a. *Harpa harpa* LINNAEUS, 1758

Rather solid with 12 to 14 broad thick and rather flat axial costae. Spire of medium height. Small pointed denticles present at the anterior end of the outer lip. Exterior cream with brown blotches and triangular lines between the costae, which are pink and crossed by bands of fine brown lines. Spire nucleus pink. The ventral parietal and columellar brown blotch divided into 3 parts. Interior usually yellowish.

7 cm. Moderately common. Indo-West Pacific. There is no positive record of this species in Qld. waters but there is a series of beach-worn specimens in the Australian Museum from North East Herald Cay in the Coral Sea, so that its presence on the Great Barrier Reef could be expected. The figured specimens are from the Philippines (3a) and Mauritius (3). *H. nobilis* RÖDING, 1798 is a synonym.

4, 4a. *Harpa amouretta* RÖDING, 1798

Slender, width little more than ½ the length, spire high. Whorls with about 15 axial ribs. Columellar and parietal callus extends on to the ventral side of the spire. Exterior cream with purple-brown blotches and axial lines and arrowhead marks between the ribs. Ribs crossed by wide purple-brown bands and thin brown lines. Parietal and columellar callus with a purple-brown blotch.

7 cm. Moderately common. Indo-West Pacific; eastern Qld. *H. minor* LAMARCK, 1816 is a synonym. The species lives in sand among corals in shallow water. *H. gracilis* BRODERIP & SOWERBY, 1829 is a closely related more slender shell from south-eastern Polynesia.

5, 5a. *Austroharpa punctata* VERCO, 1906

Rather thin, roundly ovate, spire of medium height with shouldered whorls. Protoconch mammillate with 1½ whorls and caniculate sutures. Wide spiral groove (funiculum) present on anterior end behind the anterior canal. Outer lip slightly thickened. Sculptured with 10 to 13 fine erect axial costae and fine spiral striae. Columellar and parietal callus weak and narrow. Exterior salmon-pink to pale brown with spiral rows of small brown spots or dashes, and 2 broad but indistinct darker spiral bands on the body whorl, interior white.

4 cm. Few specimens are known, all of them from S.A. One of the specimens photographed (5a) is the holotype, dredged in 20 fathoms off Newland Head, Encounter Bay, S.A. The other (5) is another specimen in the South Australian Museum from St Francis Is., S.A. Mr George Pattison of Glenelg, S.A. (personal communication) writes "I used to be on many of Dr Verco's dredging trips. In 1896 we dredged in 20 to 25 fathoms off Newland Head and Backstairs Passage, S.A., two good specimens and two broken specimens of *A. punctata*. I have one of the good specimens."

6, 6a. *Austroharpa exquisita* IREDALE, 1931

Thin, fragile, almost fusiform. Spire of medium height with shouldered whorls and a bulbous protoconch of 2 whorls. A wide spiral groove (funiculum) present on the anterior end behind the anterior canal. Outer lip slightly thickened. Sculptured with about 14 low rounded spiral ribs crossed by numerous erect, very thin and delicate axial costae (about 25 on the body whorl), with fine axial striae in the interspaces. Columella and parietal callus does not extend on to the ventral surface. Exterior cream with a few fine purple-brown spots, spire apex yellow, interior white.

3 cm. Few specimens known, all dredged or trawled in deep water. Previously recorded from off Twofold Bay, N.S.W. and off the Victorian coast. One of the figured specimens (6) is a Victorian shell from Lakes Entrance. The other specimen (6a) was dredged recently in 70 fathoms off Rottnest Is., W.A.—this being the first record of the species in that State. This shell shows some differences in shape and sculpture from the eastern specimens and examination of a larger series may eventually show that the western population is a distinct species or subspecies.

Plate 72. Harpidae (1⁷⁄₁₀× natural size)

vase shells (FAMILY VASIDAE)

THIS IS A RATHER SMALL FAMILY with only about 20 living species. Seven species, representing 3 genera, are known from Australian waters. Of these, 2 (both belonging to the genus *Vasum*) are widely distributed in the tropical Indo-West Pacific including northern Australia, 4 (all belonging to the genus *Tudicula*) are found only on our northern shores, and 1 (genus *Altivasum*) is endemic to South Australia and southern Western Australia. Little is known of the biology of any of these molluscs although it is certain they are carnivorous. Species of the genus *Vasum* live in sand and coral rubble, on coral reefs in shallow water. Species of *Tudicula*, on the other hand, live in sand among rocks at considerable depths as well as in shallow water. Vase shells are close relatives of the Chank shells (family Xancidae), in fact the 2 groups are sometimes considered to be subfamilies of a single family. The Xancidae has no known representatives in Australian waters.

Reference:

ABBOTT, R. TUCKER (1959). The family Vasidae in the Indo-Pacific. Indo-Pacific Mollusca, 1 (1) : 15—32, pls. 1-10.

1, 1a. *Vasum ceramicum* LINNAEUS, 1758

Very solid and heavy, elongately biconic to fusiform, spire high, anterior canal attenuate. Whorls strongly spined. A spiral row of large conical spines present on the shoulders. Anterior to this on the body whorl there are 3 to 5 low spiral ribs bearing smaller spines or nodules, plus a short nodulose anterior fasciole. Outer lip crenulate, bearing 5 or 6 pairs of small black teeth. Columella with 3 strong plaits and sometimes weak folds in the spaces between the plaits. Umbilicus small or closed. Exterior whitish with black-brown patches often arranged in axial and spiral bands, interior white, columella white with black-brown patches at the centre and anterior extremity.

14 cm. Moderately common. Indo-West Pacific; N.T. to northern Qld.

2, 2a. *Vasum turbinellum* LINNAEUS, 1758

Solid and heavy, biconic, spire of low to medium height. Whorls strongly spined, spire whorls and body whorl bear a spiral row of long, curved, blunt spines at the shoulders, and anterior to these on the body whorl there are 5 stout spiral ribs which bear short spines or nodules. Anterior fasciole stout and nodulose. Outer lip crenulate. Columella with 3 strong plaits and sometimes weaker folds in the spaces between the plaits. Umbilicus closed. Exterior whitish with dark brown spiral bands between the ribs. Interior and columella glazed and white. The inner edge of the outer lip with 4 to 7 squarish black-brown spots, large black-brown patches present on the columella and parietal wall.

8 cm. Common. Indo-West Pacific; N.T. to northern Qld. The shorter spire, the stout club-shaped spines, and the glazed columella and parietal wall distinguish this species from *V. ceramicum*.

3. *Altivasum flindersi* VERCO, 1914

Solid, heavy, fusiform, spire very high. Anterior canal narrow but very deep. Early spire whorls with scaly axial lamellae and strong nodulose axial folds crossed by spiral ribs. On later whorls and the body whorl, nodules become elevated open-sided spines arranged in axial and spiral rows, those on the shoulders and at the anterior end being longest. Anterior fasciole strong. Outer lip crenulate. A raised parietal shield present in adults. Columella with 3 strong broad plaits. Umbilicus wide and very deep. Exterior chalk-white pink or yellow, interior white.

18 cm. Uncommon. Rocks at depths from 10 to 120 fathoms. Backstairs Passage, S.A. to Jurien Bay, W.A. Most specimens have been trawled or taken in craypots, but scuba divers have collected living specimens recently in S.A. This is the largest and perhaps the most beautiful species in the family.

Fig. 21. Coral reefs such as the ones found in the Capricorn Group, Qld., are excellent collecting areas for species of the genus *Vasum*.
(*Photo*—Keith Gillett).

4, 4a. *Tudicula armigera* A. ADAMS, 1855

Pyriform with a tumid body and a long straight anterior canal. Spire of medium height with a large and mammillate nucleus of 1½ smooth whorls. Post-nuclear whorls and body whorl bear a spiral row of long hollow spines on the shoulders, fine spiral ribs sometimes with short spines, minute axial lamellae, and a few low rounded axial folds. At the base of the anterior canal 2 or more spiral rows of long, often curved, hollow spines present. Aperture smooth, although lip may be weakly crenulate. Parietal shield erect and solid, columella with 3 or 4 strong plaits. Exterior cream or white, sometimes finely spotted with brown, interior white.

7.5 cm. Moderately uncommon. Lives in sand from 1 to 40 fathoms. Adele Is., W.A. to Moreton Bay, Qld. The Adele Is. record is the first for W.A.

5, 5a *Tudicula spinosa* H. & A. ADAMS, 1863

Turnip-shaped with a globose body whorl and a very long straight spine-like anterior canal. Spire short, nucleus mammillate and consisting of 1 smooth whorl. Remaining spire whorls and the body whorl sculptured with fine axial striae and sometimes low axial folds, and bearing a spiral row of long hollow spines at the shoulders. About 4 spiral ribs with short hollow spines present between the shoulders and sutures, and strong spiral ribs, scaly but rarely spinose, on anterior part of body whorl and anterior canal. Aperture strongly lirate. Parietal shield erect and solid, columella with 3 strong plaits. Exterior cream often flecked with red spots, anterior canal brown. Interior white.

5 cm. Uncommon. In fine sand at depths from 9 to 100 fathoms. Rottnest Is., W.A. to Moreton Bay, Qld. The illustrated specimens come from Rottnest Is. and Onslow, W.A.

6, 6a. *Tudicula inermis* ANGAS, 1878

Turnip-shaped with a globose body whorl and a very long straight spine-like anterior canal. Spire short with a small smooth mammillate nucleus of 1 whorl. Sculptured only with fine spiral threads and very fine axial striations in the interspaces, although some specimens have very weak shoulder nodules and axial folds. Aperture strongly lirate. Parietal shield high and erect. Columella with 3 strong sharp-edged plaits. Exterior white, cream or buff with brown axial streaks or patches, interior white.

5 cm. Moderately uncommon. In sand at depths from low tide level to 40 fathoms. Houtman Abrolhos, W.A. to Darwin, N.T. Known as the "toffee-apple shell" in W.A.

7, 7a. *Tudicula rasilistoma* ABBOTT, 1959

Solid, fusiform, with a moderately high spire and a moderately long tapered anterior canal. Spire nucleus small and consists of 1½ whorls, post-nuclear and body whorls more-or-less smooth with only weak spiral threads and axial growth striae, but a single or double row of heavy nodules present on the shoulders and a spiral row of short spines around the centre of the anterior canal. Aperture smooth or weakly lirate, outer lip weakly crenulate. Parietal shield well developed, columella with 3 strong plaits. Exterior pink-cream or whitish with spiral bands and lines of brown, anterior tip dark brown. Interior white, parietal wall and columella polished, pinkish.

8 cm. Moderately uncommon. Trawled at depths from 20 to 40 fathoms. Southern Qld. and northern N.S.W.

Plate 73. Vasidae (⁸⁄₁₀× natural size)

mitres (FAMILY MITRIDAE)

MOST OF THE SEVERAL HUNDRED SPECIES OF MITRES live in tropical or temperate waters. Although a number of species inhabit the Atlantic and the west coast of tropical America, the family is most prolific in the central Indo-West Pacific region where the most colourful and intricately sculptured species occur. The coast of Queensland has a great number of beautiful species but, strangely, there are relatively few on the tropical northern shores of Western Australia. In the temperate waters of southern Australia there are several endemic species.

Mitre shells are usually long and tapered, sometimes smooth but more often delicately or strongly sculptured. They have strong plaits on the columella, a thin periostracum and no operculum. Collectors are often puzzled by the long extensible proboscis of mitres, particularly in the case of the large and common species *Mitra mitra*. When living specimens are dug from the sand this organ is often extended, and may be longer than the shell. As a rule the animal does not retract the proboscis when handled and usually dies with it still extended. Some people mistake it for a worm being eaten by the animal. The radula is located in the end of the proboscis. Another characteristic of living mitres is the foul-smelling purplish fluid they exude when being handled. Like their relatives the volutes, olives, harps and vasids, the mitres are predators or scavengers. Some are said to browse on coral polyps. Little is known of their breeding habits although many lay flask-shaped capsules under stones and have a short planktonic larval stage. Their habitats are varied. Some species live among the hard corals on coral reefs and may be found under stones or corals during the day. A large proportion of them burrow in sand. Sand-dwelling forms may be collected by "tracking" at low tide. A skindiver on the Queensland coral reefs can usually obtain a variety of small mitres by "fanning" the sand with his hands or flippers, so uncovering the shells which are otherwise invisible.

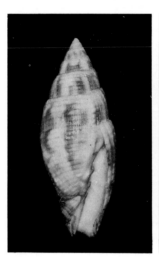

Fig. 22. *Pterygia barrywilsoni* J. CATE, 1968.
Ovate to fusiform. Whorls rounded, sculptured with numerous deeply punctate spiral grooves, and fine crowed axial cords. Anterior fasciole well developed. Columella with 4 prominent plaits. Exterior off-white, with brown-grey axial streaks, and spiral bands of yellowish blotches on the body whorl, anterior end rust coloured. Columellar plaits rust coloured.

3.89 cm. The figured specimen is the unique holotype in the collection of the W.A. Museum. It was collected at Nightcliffe, Darwin by Mrs Jo Cunningham. (*Photo*—Keith Gillett).

1, 1a. *Mitra mitra* LINNAEUS, 1758
Solid and heavy. Aperture length about ½ the total shell length. Whorls rounded and smooth except for weak spiral grooves anteriorly, anterior fasciole prominent. Outer lip with pointed denticles which become obsolete towards the posterior end. Columella and parietal wall heavily callosed, columella bears 4 or 5 strong plaits. Exterior white with spiral rows of large, irregular orange or red spots, interior white or yellowish. In life there is a thin, smooth periostracum.
18 cm. Common. Indo-West Pacific; Shark Bay, W.A. to eastern Qld.

2, 2a. *Mitra papalis* LINNAEUS, 1758
Resembles *M. mitra*, but with spines or nodules at the shoulders, and whorls sculptured by low, rough, rounded spiral ribs separated by punctate grooves.
11 cm. Uncommon. Indo-West Pacific; eastern Qld.

3, 3a. *Mitra stictica* LINK, 1807
Another species resembling *M. mitra*, but spire strongly shouldered, shoulders bearing strong pointed nodules, and whorls more-or-less flat-sided.
7 cm. Common. Indo-West Pacific; eastern Qld.

4, 4a. *Mitra eremitarum* RÖDING, 1798
Elongately fusiform, whorls rounded. Aperture length slightly more than ½ the total shell length. A finely beaded spiral rib present at the sutures, elsewhere whorls sculptured with low, rounded, spiral ribs with shallow punctate grooves between them, crossed by axial striae. Anterior fasciole obsolete. Outer lip thickened, weakly toothed. Columella with 4 or 5 strong plaits. Exterior yellow-brown with darker brown elongate patches, interior and columella cream or yellow. Periostracum in living specimens thin, dark brown.
8 cm. Common under rocks and corals. Indo-West Pacific; eastern Qld.

5. *Pterygia solida* REEVE, 1844
Solid, fusiform to ovate, sides of the spire whorls convex. Aperture length slightly more than ½ the total shell length. Whorls glossy smooth but sculptured by punctate spiral grooves which are weak at the centre of the body whorl, anterior fasciole obsolete. Outer lip sharp and smooth. Columella with 4 strong plaits. Exterior fawn with spiral rows of irregular white spots, and often a central white spiral band and fine brown spiral lines, interior and columella white.
4.5 cm. Uncommon. Eastern Qld. and N.S.W.

6, 6a. *Mitra ferruginea* LAMARCK, 1811
Rather like a stout form of *M. eremitarum*, but the spiral ribs widely spaced and very strong and the outer lip has thick prominent nodular denticles.
6 cm. Uncommon. Indo-West Pacific; eastern Qld.

7, 7a. *Pterygia nucea* GMELIN, 1791
Solid, ovate, spire low with a smooth convex outline. Whorls smooth and rounded, anterior fasciole obsolete. Aperture length about ⅓ total shell length. Outer lip sharp edged with, in adults, 7 to 10 small brown nodules on its inner edge. Columella with 5 or 6 strong plaits. Exterior white or cream with irregular brown spiral bands and spiral rows of small widely spaced brown spots, interior and columella white.
6 cm. Uncommon. Indo-West Pacific; eastern Qld.

8, 8a. *Mitra cardinalis* GMELIN, 1791
Solid, fusiform to elongate-ovate. Spire high, flat-sided and narrow, aperture length about ½ total shell length. Whorls sculptured by well spaced, prominently punctate spiral grooves, anterior fasciole low and spirally ribbed. Inner margin of outer lip weakly crenulate. Columella with 5 or 6 plaits. Exterior white or cream with spiral rows of large squarish orange-brown spots, interior and columella white.
7 cm. Moderately common. Indo-West Pacific; eastern Qld.

9, 9a. *Neocancilla papilio* LINK, 1807
Fusiform and slender. Sides of spire whorls rounded, aperture length more than ½ the total shell length. Whorls sculptured by prominent spiral ribs crossed by narrow axial grooves giving the surface a rough nodulose texture. Anterior fasciole weak, consists of a double spiral rib. Outer lip weakly crenulate. Columella with 4 or 5 plaits. Exterior whitish or grey with contrasting white and red-brown nodules, interior and columella orange.
6 cm. Common in sandy coral pools. Indo-West Pacific; eastern Qld.

Plate 74. Mitridae (1²⁄₁₀× natural size)

Mitridae (Cont.)

1, 1a. *Mitra cucumerina* LAMARCK, 1811
Stout, biconic, aperture length more than ½ the total shell length. Sculptured by thick spiral ribs which are rounded on the posterior and centre of body whorl but V-shaped in cross section anteriorly, with fine crowded axial striae between the ribs. Outer lip crenulate, columella with 3 or 4 plaits. Exterior dark red-orange or red-brown, with a central spiral band of large cream spots on the body whorl, interior pale red-brown.
 3 cm. Common. Under rocks and corals. Indo-West Pacific; eastern Qld.

2. *Mitra lugubris* SWAINSON, 1822
Elongate-ovate. Sutures impressed, shoulders low, rounded, weakly nodulose. Aperture length more than ½ total shell length. Sculptured by many rough spiral ribs crossed by fine axial cords. Anterior fasciole low, rounded. Columella calloused, with 5 or 6 plaits. Exterior brown with a broad white spiral band at the shoulders, interior and columella white or, rarely, brown.
 3 cm. Common. Indo-West Pacific; eastern Qld. and N.S.W.

3. *Mitra ticaonica* REEVE, 1844
Broad or elongately ovate, whorls rounded. Spire height variable, aperture length ½ to ⅔ total shell length. Sculptured by very shallow punctate spiral grooves which may be obsolete at the centre of the body whorl, and strong rather flat spiral ribs anteriorly. Anterior fasciole obsolete. Outer lip finely crenulate. Columella calloused, with 4 or 5 strong plaits. Exterior and interior dark tan or wine-red, columellar teeth white.
 3.5 cm. Moderately common on sandy reefs. Indo-West Pacific; Geraldton, W.A. to eastern Qld.

4, 4a. *Mitra variabilis* REEVE, 1844
Fusiform, slender, spire high, pointed. Aperture length slightly more than ½ total shell length. Whorls sculptured by spiral grooves with fine crowded axial striae between the ribs. Outer lip finely crenulate, with 4 to 5 plaits on the columella. Exterior brown with irregular white blotches, and a white central spiral band on the body whorl, interior brown.
 4 cm. Moderately common. Central Indo-West Pacific; North-West Cape, W.A. to eastern Qld.

5, 5a. *Vexillum caffrum* LINNAEUS, 1758
Fusiform and slender, anterior end up-turned. Aperture length more than ½ total shell length. Early whorls with strong axial ribs, rounded shoulders and several fine spiral striae near the sutures. Body whorl smooth except for strong flat low spiral ribs at the anterior end. Outer lip smooth but there are 6 to 10 weak lirae on the outer wall of the aperture. Columella calloused with 4 or 5 strong plaits. External colour variable, usually chocolate brown with 2 or 3 white or yellow spiral bands on the body whorl, 1 on the earlier whorls. Lip and columella dark brown, interior and collumellar teeth white.
 5 cm. Uncommon. In muddy sand. Central Indo-West Pacific; eastern Qld.

6, 6a. *Vexillum gruneri* REEVE, 1844
Fusiform to biconical, spire shouldered. Aperture length slightly more than ½ the total shell length. Whorls with a smooth polished surface, sculptured only by high angular axial ribs ending in pointed nodules just in front of the sutures, and 3 or 4 weak spiral ribs at the anterior end, anterior fasciole obsolete. Outer wall of aperture with weak lirae, columella with 5 or 6 plaits. Exterior blue-grey or brown with 1 or more darker spiral bands and several thin spiral red-brown lines, interior brown.
 3 cm. Uncommon. In sandy coral pools. Central Indo-West Pacific; eastern Qld.

7, 7a. *Vexillum plicarium* LINNAEUS, 1758
Fusiform to elongate biconic, spire shouldered. Aperture length slightly more than ½ the total shell length. Sculptured by high axial folds which form nodules at the shoulders, crossed by fine spiral grooves and, on the body whorl, nodulose ribs anteriorly. Anterior fasciole low. Weak lirae present on the outer wall of the aperture. Columella with 4 or 5 plaits. External colour variable, usually white, cream or orange with spiral bands of white or black, inner side of outer lip black and white, deep interior white.
 5 cm. Common. Sand. Indo-West Pacific; North-West Cape, W.A. to eastern Qld.

8, 8a. *Vexillum rugosum* GMELIN, 1791
Fusiform to elongate biconic, spire shouldered. Aperture length slightly more than ½ the total shell length. Sculptured by strong axial plications which form prominent nodules at the shoulders, crossed by strong spiral grooves which produce a rough surface texture. Anterior fasciole low. Weak lirae present on the outer wall of the aperture. Columella with 3 or 4 strong plaits. Exterior green-grey or blue-grey with 2 broad spiral black bands on the body whorl and 1 in the sutures. Outer lip black and white, deep interior white.
 5 cm. Uncommon. Sand. Indo-West Pacific; Exmouth Gulf, W.A. to eastern Qld. Resembles *V. plicarium*, but distinguished by the deeper spiral grooves and rough surface.

9, 9a. *Vexillum sanguisugum* LINNAEUS, 1758
Fusiform and slender, spire high, pointed. Aperture length about ½ the total shell length. Sculptured by strong axial ribs crossed by weak spiral grooves on the posterior part of the body whorl which become deeper anteriorly. Anterior fasciole low. Weak lirae on the outer wall of the aperture. Columella with 3 to 5 plaits, the posterior one strong. External colour variable, may be cream or blue-grey, with 2 spiral rows of red-orange spots (squarish or axially elongate) around the body whorl, 1 on earlier whorls. The spiral grooves contain thin brown lines. Columella brown, deep interior white.
 5 cm. Common. Sand. Indo-West Pacific; eastern Qld.

10, 10a. *Vexillum taeniatum* LAMARCK, 1811
Fusiform, very slender, anterior end constricted, drawn out and up-turned. Spire high, shouldered, aperture length slightly more than ½ total shell length. Whorls with prominent axial folds crossed by spiral grooves, and by rounded spiral riblets at the anterior end. Anterior fasciole narrow. Outer wall of aperture smooth, columella with 4 or 5 plaits. Exterior variably coloured and patterned, usually with orange and white spiral bands separated by thick black spiral lines, interior whitish, columella orange.
 7 cm. Uncommon. Sand. Indo-West Pacific; eastern Qld.

11, 11a, 11b. *Vexillum vulpecula* LINNAEUS, 1758
Fusiform, slender, anterior end up-turned. Spire high, aperture length more than ½ the total shell length. Spire whorls with strong axial ribs which become low widely spaced axial folds on the body whorl, crossed by fine spiral grooves. Spiral ribs present at the anterior end. Weak lirae sometimes present on the outer wall of the aperture. Columella bears 3 or 4 moderately strong plaits. External colour extremely variable, usually orange or orange-brown with black and white spiral bands, sometimes uniformly brown, interior white, lip brown.
 5 cm. Common. Sand. Indo-West Pacific; Port Hedland, W.A. to eastern Qld.

12, 12a. *Swainsonia casta* GMELIN, 1791
Slender and nearly cylindrical, spire high and pointed. Early spire whorls with fine axial cords and deeply punctate spiral grooves, body whorl glossy and smooth except for about 3 spiral striae at the anterior end. Columella with 5 or 6 strong plaits. Exterior white or cream, with a broad spiral band of thin, smooth and very closely adhering brown periostracum. The periostracum may be partly or entirely worn away.
 4 cm. Uncommon. Sand. Indo-West Pacific; eastern Qld.

13, 13a. *Imbricaria conularis* LAMARCK, 1811
Biconic, spire short, sharply pointed and with concave sides. Aperture narrow, aperture length more than ¾ the total shell length. Body whorl smooth, with smooth angular shoulders, anterior fasciole obsolete. Columella with 5 or 6 narrow sharp-edged plaits. Exterior blue-grey with spiral rows of squarish white spots between thin orange-brown spiral lines. Deep interior brown, white near the margin, columellar teeth white.
 2.5 cm. Common in sand between coral heads. Indo-West Pacific; eastern Qld. Commonly known as *I. conicus* SCHUMACHER, 1817.

14, 14a. *Vexillum exasperatum* GMELIN, 1791
Elongate-ovate, constricted anteriorly. Spire shouldered, aperture length less than ½ total shell length. Sculptured by axial ribs crossed by numerous fine spiral grooves. Many lirae present on the outer wall of the aperture. Columella bears 4 strong plaits. Exterior cream or white, sometimes 2 brown patches present on the outer wall of the aperture, otherwise interior and columella white.
 2.5 cm. Common in sand. Indo-West Pacific; eastern Qld. The strength and number of axial ribs is variable.

Plate 75. Mitridae (1⁴⁄₁₀× natural size)

Mitridae (Cont.)

1. *Eumitra chalybeia* REEVE, 1844

Elongately fusiform. Spire high and tapered to a point, aperture length slightly less than ½ the total shell length. Sides of spire whorls slightly convex, whorls smooth except for a few very fine granular spiral striae anterior to the sutures. Anterior fasciole low. Columella with 3 or 4 plaits. Exterior blue-grey or brown-grey with many fine spiral brown lines, interior brown-grey, columella light brown.

6 cm. Common. Among seaweeds and under stones. Southern W.A. from about Albany to Shark Bay.

2. *Eumitra glabra* SWAINSON, 1821

Elongately fusiform. Spire very high, tapered to a point, aperture length about ⅓ total shell length. Spire whorls with rather flat sides, whorls sculptured with fine, well spaced, finely punctate spiral grooves and fine longitudinal growth striae. Anterior fasciole low. Exterior fawn or light brown with slightly darker brown in the spiral grooves, interior pale, columella pale brown or white.

9 cm. Moderately common. N.S.W. to Fremantle, W.A. The surface is covered by a dark brown periostracum in life. This species is easily confused with *E. chalybeia* but may be distinguished by the higher spire and brown coloration.

3. *Eumitra badia* REEVE, 1845

Elongately fusiform. Spire high and tapered to a point, aperture length about ½ the total shell length, outer lip curved. Spire whorls with convex sides, whorls polished smooth but sculptured with minute, punctate, spiral striae crossed by equally minute axial striae. 8 to 10 well spaced spiral grooves present at the anterior end of the body whorl. Anterior fasciole low. Columella with 3 or 4 rather weak plaits. Exterior dark brown, interior and columella lighter brown.

4 cm. Common. N.S.W. to Fremantle, W.A. The N.S.W. form (eg. 3) is more elongate than the southern and western form and is sometimes considered a distinct species with the name *E. rhodia* REEVE, 1845.

4. *Eumitra nigra* GMELIN, 1791

Elongately fusiform. Spire high and tapered to a sharp point, aperture length about ½ the total shell length, outer lip rather straight. Spire whorls with slightly convex sides, and fine spiral and longitudinal striae which are microscopically punctate. Body whorl with rather coarser spiral striae. Anterior fasciole obsolete. Columella straight and bears 4 or 5 plaits. Exterior uniformly dark brown with fine darker brown spiral lines, interior grey-brown, columella pink-brown.

7 cm. Moderately common. Pacific, N.S.W. to S.A. This species is known in Australian literature as *E. melaniana* LAMARCK, 1811 (= *E. contermina* IREDALE, 1936) but Cernohorsky (1967) considers the Australian form conspecific with the tropical *E. nigra*.

5. *Eumitra cookii* SOWERBY, 1874

Like a slender form of *E. badia*, but distinguished by a broad, pale spiral band around the centre of the body whorl.

3 cm. Uncommon. Southern Qld. and northern N.S.W.

6. *Eumitra australis* SWAINSON, 1822

Elongately fusiform. Spire of medium height, tapered to a point, aperture length slightly less than ½ the total shell length, outer lip curved. Spire whorls with convex sides, whorls polished, smooth except for fine growth striae, 2 or 3 fine spiral striae in front of the sutures and several others at the anterior end of the body whorl. Anterior fasciole low. Columella with 4 thick plaits which may continue around as low spiral ridges on the anterior fasciole. Exterior brown with a broad cream central spiral band, interior and columellar plaits white.

4 cm. Moderately common. Vic., Tas. and S.A.

7. *Austromitra bucklandi* GABRIEL, 1961

Fusiform, with strongly shouldered spire whorls. Aperture length about ½ the total shell length. Sculptured with fine spiral striae, several fine spiral grooves at the anterior end (2 rather deep), and with strong axial ribs which become obsolete towards the anterior end of the body whorl. Anterior fasciole low. Columella with 4 plaits. Exterior cream to light orange-brown with a broad interrupted brown spiral band around the centre of the body whorl, interior white or fawn, columella white or pale mauve.

2 cm. Moderately common. Dredged off N.S.W. and Vic.

8. *Pusia patriarchalis* GMELIN, 1791

Broad and ovate to biconic. Whorls strongly shouldered, with heavy pointed nodules on the shoulders, sculptured by punctate spiral grooves and spiral rows of rounded nodules anteriorly. Anterior fasciole low. Columella with about 4 strong plaits. Exterior white or cream with a broad central spiral band of orange or red-brown, interior and columella white or cream.

3 cm. Uncommon. Indo-West Pacific; eastern Qld.

9. *Strigatella litterata* LAMARCK, 1811

Heavy, solid, ovate. Aperture length more than ½ total shell length. Whorls smooth except for a few indistinct punctate spiral grooves at the anterior end and on the spire whorls. The inner side of the outer lip calloused, columella with 4 or 5 strong plaits. Exterior white or cream with fine brown spiral lines, irregular axial brown streaks at the sutures and at the anterior end, and a central spiral band of irregular axial brown blotches. Interior and columella white.

3 cm. Common on shallow reefs. Indo-West Pacific; eastern Qld. to N.S.W.

10. *Strigatella scutulata* GMELIN, 1791

Solid, heavy, elongate-ovate. Aperture length more than ½ the total shell length. Body whorl smooth except for a few spiral cords anteriorly and posteriorly near the sutures, spire whorls with spiral ribs. Anterior fasciole short and low. Outer lip sinuous, smooth or sometimes weakly crenulate. Columella with 4 or 5 strong plaits. Exterior dark brown, sometimes with a broad yellow spiral band near the sutures.

5 cm. Common on shallow sandy reefs. Indo-West Pacific; Cape Naturaliste, W.A. to N.S.W. The illustrated specimen is of the banded form from W.A. which is sometimes known as *S. amphorella* LAMARCK, 1811.

11. *Strigatella crassa,* SWAINSON, 1822

Solid, heavy, elongate-ovate, sutures deeply impressed. Aperture length slightly more than ½ the total shell length. Whorls sculptured with many spiral ribs, those at the posterior end flattened, those at the anterior end rounded. Anterior fasciole low. Outer lip weakly crenulate, columella with 5 or 6 strong plaits. Exterior dark brown with a broad yellowish spiral band near the sutures, interior and columella white.

5 cm. Common on shallow reefs. Central Indo-West Pacific; eastern Qld.

12. *Cancilla nodostaminea* HEDLEY, 1912

Slender and fusiform, spire whorls slightly shouldered with convex sides. Aperture length about ½ the total shell length. Whorls with a rough cancellate structure formed by raised longitudinal and spiral ribs with low nodules at the intersections. Columella with 3 to 5 plaits. Exterior uniformly cream to light brown, interior and columella pale mauve.

2.5 cm. Uncommon. Dredged off N.S.W. This species resembles *C. strangei* ANGAS, 1867 which occurs in the same area, but in that species the spiral sculpture predominates.

13. *Cancilla filaris* LINNAEUS, 1771

Fusiform to elongate-ovate, spire whorls with low rounded shoulders. Aperture length slightly more than ½ the total shell length. Whorls sculptured by 10 to 12 nodular spiral ribs with finer nodular spiral cords in the interspaces, crossed by fine axial grooves. Anterior fasciole low. Columella with 4 or 5 plaits. Exterior light brown or whitish, spiral ribs dark brown, interior and columella white or blue-white.

3 cm. Common in sandy coral pools. Indo-West Pacific; eastern Qld.

14, 14a. *Cancilla interlirata* REEVE, 1844

Fusiform, very slender, spire high and pointed, anterior canal drawn out, re-curved and slightly up-turned. Aperture length slightly more than ½ the total shell length. Sculptured by high narrow sharp spiral ribs (about 12 to 14 on body whorl) with less prominent spiral riblets between, and by delicate axial riblets. Anterior fasciole obsolete. Outer lip weakly crenulate, columella with 4 or 5 plaits. Exterior white with spiral bands of light brown patches, spiral ribs brown (sometimes not continuously), interior white.

4 cm. Uncommon in deep water. Indo-West Pacific; recently dredged off Onslow, W.A. (Qld.?).

Plate 76. Mitridae (1⁷⁄₁₀× natural size)

Plate 77. *Amoria canaliculata* McCoy, 1869, Qld. (*Photo*—Don Byrne).

volutes (FAMILY VOLUTIDAE)

THE VOLUTES ARE PREDACEOUS GASTROPODS. More than the 200 living species are known. The family is best represented in the warm seas of the world but many temperate species also occur in the Atlantic, and off southern Australia, New Zealand and Japan. There are a few species in the Antarctic. Australia is usually considered to be the home of volutes. The numerous Australian species (about 67) are often more brightly coloured and patterned than their cousins elsewhere.

Like the cowries and cones, volute shells are favourites with collectors. Apart from their beautiful colours, they have graceful shapes and glossy surfaces and many of them are so hard to obtain that they excite our collecting instincts. Some species still bring high prices on the shell market but in recent years the habitats of a surprising number have been discovered. Examples of once rare Australian species which have become more easily available are *Amoria macandrewi*, *Ternivoluta studeri* and *Volutoconus grossi*. Examples of species still known only from a few specimens are *Paramoria weaveri*, *Volutoconus coniformis*, *Notopeplum translucidum*, *Nannomoria amicula*, *Amoria spenceriana* and *Notovoluta perplicata*.

Volutes lay their eggs in tough almost cylindrical egg masses, and in all species which have been studied the young hatch directly from the eggs as miniature crawling snails already equipped with a protective shell, i.e. there is no free-swimming larval stage. The egg case of the large baler shells *(Melo)* is shaped like a pineapple and consists of a number of individual capsules, each like a pineapple segment (Fig. 29, page 138). Only 1 egg develops in each capsule, and when the babies hatch they have a shell which is already very like that of their parents.

There have been reports that female balers, like cowries, "sit" on their egg-masses. Certainly the females show a strong "maternal instinct" towards their eggs during laying. Late one Sunday afternoon in early December 1969 one of the authors was given a female *M. miltonis* and a half-completed egg-mass collected that morning in Cockburn Sound, W.A. When found the baler was sitting on top of the egg-mass which was like an open-ended cylinder about 15 cm high and 13 cm diameter. It contained approximately 50 capsules and was attached to a dead razor clam *(Pinna)*.

The mother baler and the egg-mass were kept in a bucket and covered with a wet towel all day. The author took possession of them that evening and put them into a 15-gallon tub of sea water without aeration.

The next morning both the baler and the egg-mass were transferred to a large shallow aquarium pond with running sea water at the W.A. Marine Research Laboratory at Waterman, W.A. The mother baler quickly recovered and became very active. After crawling about for nearly an hour she made contact with her egg-mass on the bottom of the pond. She immediately climbed back on top of it and remained firmly attached to it for another 10 days. During this time she added another 20 capsules to the egg-mass. On the tenth day she left the egg-mass, which was by then completed and sealed off at the top, and climbed out of the pond to crash on to the concrete floor. This seemed to indicate that the mother had no further interest in her eggs and was moving to seek food or shelter somewhere else. At the time of writing (January 31, 1970) none of the embryos had reached the hatching stage but all, including those in the capsules added in the aquarium, were well advanced shelled snails.

As a measure of the effectiveness of this form of larval development the case may be cited of a large *Melo amphora* egg-mass trawled in Shark Bay in 1967 and presented to the Western Australian Museum by Mr Alan Nichol of Carnarvon. The whole egg-mass was 38 cm high and contained 202 individual capsules. Development was far advanced and in every capsule there was a juvenile baler shell on the point of hatching. In other words every egg that the mother had laid developed successfully to the hatching stage. Presumably there must be a high mortality immediately after hatching due to predation and other factors, but at least the embryos receive the best possible nutrition and protection up to that stage.

The lack of a free-swimming larval stage in volutes means that dispersal of the animals away from their place of birth depends on the locomotory ability of the crawling snail. Probably for this reason many species of volutes are found in very restricted areas. For example, *Amoria macandrewi* has only been found around Barrow and the Monte Bello Islands, W.A. and its relative *A. ellioti* is known only from a small area near Port Hedland about 200 miles away. No volutes have widespread distributions throughout the Indo-West Pacific region as have many cowry species.

Some volutes, including the Queensland *Cymbiolacca pulchra* species complex, are highly polytypic i.e. several or many distinctive localized populations occur, and it is tempting for taxonomists to give each of these local variants a distinct species or subspecies name. However, the causes of this variation are not yet understood and a conservative attitude toward the taxonomy of such polytypic species seems preferable at this stage. It has been suggested that, because of the lack of a free-swimming larval dispersal stage, localized populations of shallow water volutes may be "inbreeding" and that this may result in the variation of shell characters

between localities. Shell characters may also vary with environmental condition to some extent. Detailed biological studies will be needed to clarify these problems.

Most volutes burrow in sand. They have a large, often very colourful foot, and can crawl more quickly than most other snails. They also have relatively efficient respiratory and nervous systems. All these adaptations help them as predators, for volutes catch and eat other small creatures on the sea bed. Some kinds of volutes may be found on intertidal sand flats, particularly in northern Australia where there is a large tidal range. They tend to prefer clean, well washed and sorted sand. As the tide goes out they bury and their presence is impossible to detect, but just as the tide turns to flow in again they sometimes pop up and begin crawling on the sand before the water reaches them. At that time the lucky collector can pick them up alive without even getting his feet wet. There are probably few greater shell collecting thrills than volute collecting on the vast sand flats of north Western Australia where a variety of "rare" species may be collected in this way on the odd day when the conditions are right. But unfortunately many species are only found in deep water and can be collected alive only by dredging, trawling or diving. Most of the species still rare in collections are such deep water forms.

References:

The generic names used for volutes in this book follow the list published by Mr Cliff Weaver and the Hawaiian Malacological Society in 1964.
An unusual recent publication of special interest is "Multiform Australian Volutes" published privately in Sydney by Mr Frank Abbottsmith in 1969. Although not intended to be a scientific treatise on the family, it is nonetheless the first comprehensive treatment of all the volute species in this country, and contains a vast amount of information on the variation, distribution and habitats of Australian volutes. Collectors will find it a most useful document. Other, less comprehensive works on Australian volutes are scattered through a number of journals too numerous to be listed.

Plate 78. *Melo amphora* SOLANDER, 1786, Qld. This unusual photograph shows a juvenile baler devouring the volute, *Zebramoria zebra* LEACH, 1814.
(*Photo*—Don Byrne).

Plate 79. *Cymbiolena magnifica* GEBAUER, 1802.

Left—Egg-mass of *C. magnifica,* at a depth of 40 feet, off Bare Island, Sydney. (*Photo*—Walter Deas).
Middle—*C. magnifica,* in natural habitat, off Bare Island, Botany Bay, depth 35 feet, (*Photo*—Neville Coleman).
Right—*C. magnifica,* with animal partially extended, off Bare Island, Botany Bay, depth 35 feet. (*Photo*—Neville Coleman).

1. *Cymbiolena magnifica* GEBAUER, 1802
Large and ventricose with a short pointed spire and a smooth rounded but not bulbous protoconch. Aperture length more than ¾ of the total length. Shoulders slightly angled with large low nodules. Adult whorls cream or pale brown, with irregular diffuse zigzag brown lines demarcating ragged triangular cream patches of varying size. 4 bands of darker brown markings encircle the body whorl. With 4 orange columellar plaits, interior orange.

35 cm. Not uncommonly trawled off eastern Australia between southern Qld. and Vic. The illustrated specimen is a juvenile.

Fig. 23. *Notopeplum translucidum* VERCO, 1896.
Thin, glossy smooth, with a high spire, a simple protoconch, and a fairly large, swollen body whorl. Anterior canal broad and shallow. With 4 columellar plaits. Pale creamy-grey with a narrow red-brown pre-sutural line, axial lines of elongated or zigzag spots and 2 darker spiral bands.

4 cm. Few specimens known. Deep water off S.A. The figure is a drawing copied from the original illustration by Verco.

2, 2a. *Ericusa papillosa* SWAINSON, 1822
Solid and heavy with a thickened lip, rounded shoulders and a slightly papilliform protoconch. Aperture length about ¾ the total length, with 4 strong columellar plaits. Exterior pale yellow with irregular diffuse brown splashes which form 2 darker brown bands encircling the body whorl. Columella and interior pale yellow.

15 cm. Uncommon. Trawled, southern Qld. and N.S.W. to the south Coast of W.A. One of the specimens illustrated (2) is the typical form from S.A., the other (2a) is from southern N.S.W. and has the axial ridges characteristic of specimens from that region. The axially ridged form is found from southern Qld. to the east coast of Tas. and is recognized as a subspecies, *E. papillosa kenyoniana* BRAZIER, 1898, by some authors.

3. *Livonia mammilla* SOWERBY, 1844
Large, ventricose, with a solid bulbous yellow protoconch, rounded shoulders, a reflected lip in adults, and with 3 low and broad columellar plaits. Shoulders lacking colour pattern but sculptured by fine spiral striae. Body whorl cream with 2 or 3 broad bands of irregular zigzag light brown markings. Interior and columella glossy cream or orange.

30 cm. Not uncommonly trawled off eastern Australia from southern Qld. to Vic. and Tas.

4. *Livonia roadnightae* McCOY, 1881
Large, solid, fusiform, with a thickened outer lip and a bulbous protoconch. Aperture length about ¾ the total length. Body whorl smooth except for fine spiral ribs and low axial ridges on the shoulders, ridges stronger on the spire whorls. Exterior cream to light brown with dark brown irregular zigzag lines.

20 cm. An uncommon species trawled in deep water from southern N.S.W. and Vic. to Rottnest Is., W.A. Rarely taken alive.

5, 5a, 5b. *Cymbiolista hunteri* IREDALE, 1931
Rather thin, cylindrical, with a low spire and nodulose angulate shoulders. Protoconch very small, apical whorls about 4. Whorls smooth and glossy. Columella with 4 very strong plaits with undercut sides. Lip flared, deeply notched at the posterior as well as the anterior end. Colour variable, exterior usually fawn or buff with strong brown sub-sutural lines and irregular brown axial dashes or continuous brown zigzag lines on the body whorl, and with 3 spiral bands of glazed purple-brown blotches. Columella and aperture rich rusty orange. Some shells tinged green, with orange or orange-edged axial lines (e.g. 5b).

18 cm. Moderately common. Trawled off southern Qld. and N.S.W. The most common colour form is represented by (5, 5a) both specimens from N.S.W., (5b) is a beautiful colour form from Caloundra, Qld.

6. *Mesericusa sowerbyi* KIENER, 1839
see also pages 138, 139, plate 92 (5).
Moderately large, fusiform with a long pointed spire and rounded shoulders. Aperture length about ⅔ the total length. Protoconch small but papillose. With 3 thick but low columellar plaits. Pale creamy-orange or flesh-coloured with dark, chestnut-brown, zigzag markings and a faint, pale band encircling the middle of the body whorl.

20 cm. Not uncommonly trawled off south-eastern Australia between Tas., Vic. and N.S.W. The shell is variable.

Plate 80. Volutidae (⁹⁄₁₀ × natural size) ⏵

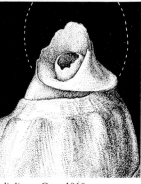

Fig. 24. Protoconch of *Cottonia nodiplicata* Cox, 1910.
Left: The large and bulbous protoconch intact in a juvenile specimen.
Right: Protoconch broken off as in most adult shells. Note the shelly internal plug which seals the broken end and the similarity of this protoconch to those of the fossil genus *Pterospira* (see Fig. 25, this page).

1. *Cottonia nodiplicata* Cox, 1910

Adult shell large with a thickened lip (but most specimens in collections are juvenile and have a thin, broken lip). Shoulders low but angular and bear a row of strong nodules. With 2 or 3 low columellar plaits. Orange with a pale band around the shoulders and a thin pale band around the middle of the body whorl. Protoconch large, bulbous and very fragile (Fig. 24, left), usually broken off early in life (Fig. 24, right).

36 cm. Uncommon. S.A. to Jurien Bay, W.A. Specimens have been collected alive by skin divers off the south-west coast but most specimens are from deep water (to 100 fathoms).

2, 2a. *Ericusa fulgetrum* SOWERBY, 1825

Solid, fusiform, glossy, smooth, high spired with rather broad rounded shoulders. With 3 weak columellar plaits and a rounded papilliform protoconch. Colour and pattern extremely variable, exterior ground colour usually cream or flesh-tinted overlain by brown lines, bands, spots or blotches. A common form (e.g. 2a) has broad and irregular undulating axial lines, sometimes crossed by 1 or 2 spiral bands, spiral lines, or spiral rows of spots, blotches or crescent-shaped markings. The axial lines sometimes lacking so that the shell lacks markings altogether or has only the spiral bands, lines etc.

15 cm. Moderately uncommon. S.A. and the south coast of W.A. Verco (1912) gave various names to many of the colour forms but these have no taxonomic status. A deep water rather slender flesh-coloured form marked with only a few small brown spots (e.g. 2) was named *E. orca* by Cotton (1951) but this form is also probably only a variant of *E. fulgetrum*.

3. *Notovoluta verconis* TATE, 1892

Small, solid and fusiform. Aperture length a little more than ½ the total length. Shoulders angular and bear strong elongate nodules. White with brown spiral zigzag lines.

4 cm. Uncommon. In deep water S.A. Named after Sir Joseph Verco who contributed so much to our knowledge of the molluscs of S.A. and southern W.A.

4. *Notovoluta kreuslerae* ANGAS, 1865

Solid, high spired. Aperture length a little more than ½ total length. Shoulders sharply angled and bear low nodules. Columella with 4 or 5 plaits. Adult whorls orange-brown with pale yellow triangular patches of varying size, sometimes with 2 slightly darker brown spiral bands. Interior pale yellow.

10 cm. Very uncommon off S.A. and the south coast of W.A. *N. occidua* COTTON, 1946 is like a slender, less strongly shouldered form of this, trawled from the southern coast of W.A. Another member of this genus is *N. perplicata* from Qld., see Page 133 (3).

5. *Amoria exoptanda* REEVE, 1849

Broad, solid, with angular but smooth shoulders, a short spire and 4 strong columellar plaits. Body whorl pale orange with 2 darker spiral bands, overlain by fine chestnut zigzag lines. Dark chestnut sutural blotches present, and short dark chestnut lines arising in the sutures and running a short distance forward over the body whorl. Interior rich orange.

15 cm. Very uncommon. S.A. and the south coast of W.A. Divers have recently collected living specimens at night off the S.A. coast.

6. *Paramoria guntheri* SMITH, 1886

Rather solid, with a rounded but projecting spire and broad, slightly angular shoulders bearing well spaced nodules. White, adult whorls marked by thin, light brown undulating axial lines which may be crossed by 2 spiral lines or bands of very short and crowded straight axial dashes.

5 cm. Uncommon. S.A. and the south coast of W.A. The form with 2 spiral lines or bands (e.g. 6) was once known as *adcocki* TATE, 1888 but there is now little doubt that it is conspecific with the form with only wavy axial lines (not illustrated).

P. weaveri McMICHAEL, 1961 is like this, but smaller with a short spire, angular shoulders and more zigzagged lines, known only from 2 specimens trawled in 70 to 80 fathoms off the Houtman Abrolhos, W.A.

7. *Amorena sclateri* COX, 1869

Glossy, smooth and solid, tending to be angular at the shoulders. Spire with slightly concave sides, apical whorls slightly swollen and a little off centre. 4 or 5 strong columellar plaits present, the more posterior ones sometimes bifid. Exterior and interior porcellaneous white or pale yellow.

10 cm. Uncommon. Recorded only from Bass Strait and northern Tas. Once known as *kingi* Cox, 1871. This species is considered by some to be only an "albino" form of *A. undulata*.

8, 8a. *Amorena undulata* LAMARCK, 1804

Solid, smooth, rather broad with rounded shoulders and a pointed spire. Aperture length more than ¾ the total length, columella with 4 strong plaits (sometimes additional weak plications present between the main plaits). Exterior cream or rufous red, usually with wavy zigzag axial lines on all except the 3 apical whorls, the wavy lines widely spaced or crowded, continuous or interrupted. Some specimens lack the lines (8). Interior and columella pale or dark orange.

10 cm. Common. Southern Qld. to Tas. and S.A. The colour and pattern of this species is extremely variable.

9. *Amorena benthalis* McMICHAEL, 1964

Small and robust, glossy smooth. Spire short, apex bluntly rounded, sutures glazed, body whorl weakly shouldered. Anterior fasciole weak, columella with 4 strong plaits. Exterior cream or fawn, with an ill-defined brown pre-sutural band, 2 spiral bands of brown blotches on the body whorl, and numerous fine undulating red-brown axial lines. Anterior end suffused with brown. Interior white to orange.

4 cm. Uncommon. Trawled at depths of about 100 fathoms off southern Qld. and northern N.S.W. It resembles *A. undulata*, but in the latter species the axial lines are more acutely undulated and the spiral bands of blotches are not present. *A. undulata* also occurs in southern Qld. waters but not so deep as *A. benthalis*.

Fig. 25. *Pterospira hannafordi* McCOY, 1866. A fossil shell of Miocene age from Fossil Beach, Balcombe Bay, Victoria. This is one of many fossil volute species found in the rich Tertiary beds of south-eastern Australia. Shell length 16.5 cm. (*Photo*—Keith Gillett).

Plate 81. Volutidae (⁸⁄₁₀ × natural size)

1   2   2a

3   4   5   6

7   8   8a   9

Volutidae (Cont.)

1, 1a. *Cymbiolacca wisemani* BRAZIER, 1870
(See also colour plate of live animal, page 8, plate 3).
Solid, like *C. complexa* in shape but pale orange-cream, with triangular patches of white and 3 bands of darker orange or chestnut. Sometimes minute dark orange or chestnut spots present. Pre-sutural lines indistinct or absent, but pre-sutural blotches sometimes present. Interior white.
    8 cm. Uncommon. North Qld. *C. randalli* STOKES, 1961 (1a) from north of Cairns, Qld. is probably only a form of *C. wisemani* in which the spots are developed as elongate dashes.

2, 2a. *Cymbiolacca cracenta* MCMICHAEL, 1963
Slender to almost cylindrical, with weakly angulated shoulders bearing short, sharp, backward-pointing spines. With 4 strong columellar plaits. Exterior ground colour pinkish with 3 broad spiral bands of dark russet, numerous flecks of lighter russet, and a sprinkling of fine black spots. Aperture white, bordered with orange-brown.
    8 cm. Uncommon. Localized between Townsville and Bowen, Qld. This may be only a narrow reddish northern form of *C. complexa.*

3, 3a. *Cymbiolacca peristicta* MCMICHAEL, 1963
Solid, smooth, with broad angulate shoulders bearing short sharp spines which continue as strong axial folds on to the posterior part of the body whorl. Columella with 4 strong plaits. Pink-white, with numerous very dark brown or black spots of varying size scattered over the shell, and adjacent to the black spots are pale pink to brown triangular patches. Short dark brown axial pre-sutural lines present. Aperture pink, lip bordered with orange.
    7 cm. Moderately common at the restricted locality. Swain Reefs, Qld. The shell differs from *C. pulchra* in possessing black spots all over the shell, and from other *Cymbiolacca* species in the strength and character of the spots. In view of the extreme colour variation in this genus, the status of this species needs further consideration.

4, 4a, 4b. *Cymbiolacca pulchra* SOWERBY, 1825
Rather fusiform with shoulders bearing backward-pointing spines. With 3 apical whorls which bear white axial ridges, and 3 other whorls in the adult shell. Colour extremely variable; in the most common form (4) the ground colour is pink-orange with triangular white patches of varying size, and 3 bands of irregularly shaped dark brown spots. Pre-sutural axial lines reach forward nearly to the angle of the shoulder. Interior white with an orange-pink margin around the outer lip.
    8.5 cm. Moderately common. Central and southern Qld. Each local population of this species has its characteristic form, some of which have been named. *C. woolacottae* MCMICHAEL, 1958 and *C. perryi* OSTERGAARD and SUMMERS, 1957 are now considered to be synonyms of *C. pulchra.* The *"woolacottae"* form (4a) is short and squat with strong angular shoulders and strong spines, the colour is pink and the pre-sutural lines are indistinct. The *"perryi"* form (4b) has a similar shape but bears no spots and the white patches on the surface are very large.

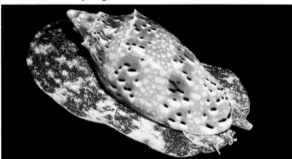

Plate 82. *Cymbiolacca pulchra* SOWERBY, 1825, Qld.
(*Photo*—Don Byrne).

5, 5a. *Cymbiolacca complexa* IREDALE, 1924
Solid, with subangulate shoulders bearing short sharp spines or nodules. Apical whorls with a few weak nodules. Orange flecked with faint triangular patches, and usually 2 or 3 darker orange or brownish spiral bands, the whole surface spotted with irregular dark brown dots. Dark brown pre-sutural axial lines present which do not reach the shoulders. Interior white.
    10 cm. Moderately common. Trawled off the eastern coast from central Qld. to northern N.S.W. at depths to at least 100 fathoms.

An extremely variable species. Some forms lack the shoulder spines or nodules. A form from Hervey Bay, Qld. (5a) is like typical *C. complexa* but is narrower, darker, has larger spots and less angular shoulders. This form has been given subspecific rank as *C. complexa nielseni* MCMICHAEL, 1959 but the distinction seems doubtful.

6. *Pseudocymbiola provocationis* MCMICHAEL, 1961
Rather fusiform with a high spire and upright pointed spines on the shoulders. Protoconch rounded and almost smooth, consists of about 2 whorls. With about 3 adult whorls. Aperture length about ¾ the total shell length. Columella with 4 plaits, inner side of the outer lip calloused. Orange-fawn with faint orange patches, a few scattered orange-brown dots and series of short vertical pre-sutural orange-brown lines.
    4.5 cm. Only a few specimens known. Trawled off Ulladulla and Port Kembla, N.S.W. at depths of about 70 fathoms. This recently described species bears a resemblance to species of the genus *Cymbiolacca*, especially *C. complexa*, but its real affinities are said to lie with a group of fossil species found in Tertiary rocks of southern Australia. However, Abbottsmith (1969) suggests that *provocationis* may be an extreme end-of-range form of *complexa* because intermediate forms are now known from northern N.S.W. Dr McMichael's choice of name for his new species would seem to have been apt.

7, 7a, 7b. *Lyria mitraeformis* LAMARCK, 1811
Solid, fusiform to ovate and strongly ribbed with broad axial plications. The 2 most anterior columellar plaits strong and thick, with a series of finer plications behind these to about half way along the columella. Exterior cream with numerous very thin interrupted spiral lines and 3 spiral bands of streaky purple-brown blotches. Interior white or yellow.
    5 cm. Not uncommon in shallow water. Vic. and Tas. to the south coast of W.A.. W.A. specimens (e.g. 7b) tend to be broader and less strongly ribbed than those from the east. This is the form recently redescribed by Weaver and Du Pont (1968) as *L. grangeri* SOWERBY, 1900. There is little doubt that this is indeed Sowerby's species but the north W.A. locality given by Weaver and Du Pont seems to be erroneous, for as far as we can determine specimens of this kind are found only on the south coast between Esperance and Cape Leeuwin. The relationship between the smooth and tumid western form and the eastern *mitraeformis* has yet to be established but we propose to treat them as 1 species for the time being.

8. *Lyria deliciosa* MONTROUZIER, 1859
Short, solid, lip thickened, shoulders rounded, spire apex sharply pointed. Early spire whorls ribbed, penultimate and body whorl smooth. Columella with 3 strong plaits anteriorly, in front of a series of fine striations. Yellow or yellow-fawn, with interrupted axial streaks of white blotches and numerous interrupted thin spiral lines. Spire apex violet, interior white.
    3.5 cm. Uncommon. Central Qld. to northern N.S.W. The Australian form of this species (e.g. 8) was named *L. howensis* IREDALE, 1937 (type locality Lord Howe Island) and is narrower and smaller than the typical form from New Caledonia. *L. opposita* IREDALE, 1937 is a synonym.

9. *Lyria nucleus* LAMARCK, 1811
Small, solid and ovate, with strong axial ribs or plications on all the whorls. Body whorl also has spiral cords at the anterior end. Columella with 2 strong plaits at the anterior end in front of a third but much smaller plait, and then a few small lirae. A weak parietal nodule or callus sometimes present beside the posterior canal. Exterior fawn or beige with purple or brown blotches and spiral rows of dots or dashes, aperture lilac, spire apex dark brown or purple-brown.
    2 cm. Uncommon. Norfolk Is., Kermadec Is. and northern N.S.W. The Australian form (as illustrated) is sometimes known as *L. pattersonia* PERRY, 1811 but that name is generally regarded as a synonym. Iredale (1940) gave yet another name to the Australian form which he called *L. peroniana.*

10. *Lyreneta laseroni* IREDALE, 1937
Thin, cylindrical, smooth, with a very small and low spire, and rather squared shoulders on the body whorl. Columella with 2 strong plaits anteriorly, and 1 or more very small plaits behind them. Exterior of freshly dead shells chocolate brown with 2 paler spiral bands and numerous dark brown spiral lines. Columellar teeth and lip white, deep interior violet.
    3 cm. Very uncommon. N.S.W. Living specimens are not yet known. Originally described as *Voluta brazieri* Cox, 1873 but that name was preoccupied.

Plate 83. Volutidae (natural size)

Volutidae (Cont.)

1, 1a, 1b, 1c, 1d. *Amoria damoni* GRAY, 1864
A variable wide-ranging species. 3 geographically separated sub-species may be recognized although the correctness of this interpretation of the species complex needs verification.

(i) *A. damoni damoni* GRAY, 1864 (1a, 1b).
Slender, smooth, shoulders low and rounded, spire rather high and pointed. With 4½ apical whorls and 3½ adult whorls. Columella with 4 strong plaits. Apical whorls grey-brown; adult whorls creamy-white or fawn overlain by a reticulate pattern of brown, irregular zigzag axial lines. Zigzag lines usually crowded in 3 spiral bands on the body whorl, within which are distinct white triangular areas. Short brown axial pre-sutural lines and a rusty brown sutural band present on the adult whorls of the spire. Lip margin and columella white, deep interior brown.
13 cm. Moderately common. North-West Cape to the North Kimberley region of W.A. Even within this area there is much variation. Probably the best known colour form is that from Broome (1a). Further north the shells are more slender, with more distinct and more widely spaced markings (e.g. (1b) from Pender Bay). Shells from the Dampier Archipelago and Exmouth Gulf area (not illustrated) are quite dark and the spiral bands are very prominent. This form was once known as *A. reticulata* REEVE, 1844.

(ii) *A. damoni reevei* SOWERBY, 1864 (1).
Solid, swollen and broad at the shoulder, with a rather short spire. Background colour pale tan to cream and the pale triangular areas usually large and prominent.
10 cm. Uncommon. Central west coast of W.A. between Fremantle and Carnarvon. The illustrated specimen was live-taken at Fremantle. Some authors regard this as a distinct species.

(iii) *A. damoni keatsiana* LUDBROOK, 1953 (1c, 1d).
A slender form characterized by rather straight and widely spaced axial lines which do not form a dense reticulate pattern.
9 cm. Uncommon. Kimberley, W.A. to Cooktown, Qld. Some authors regard this also as a distinct species. The recent report (Abbottsmith, 1969) that the range of this form extends into the range of nominate *A. damoni* in the Kimberley area lends support to this argument. (1d) is a shell from Cooktown, Qld., (1c) is a shell from Thursday Is.

2, 2a. *Amoria praetexta* REEVE, 1849
Small and slender, without shoulders, rather solid but with a sharp lip, 4 pink-grey apical whorls and 2 or 2½ adult whorls. Columella with 4 plaits. Adult whorls covered by a densely reticulate pattern of light brown zigzag lines demarcating white triangular patches varying in size from small to minute, and with 2 or 3 spiral bands of irregular dark brown blotches, and dark brown pre-sutural blotches. Interior brown, columella and the inner margin of the lip white.
7 cm. Uncommon. W.A. between North-West Cape and the North Kimberley. Shells from the southern end of the range (e.g. (2) from Onslow) tend to be darker and more finely patterned than shells from further north (e.g. (2a) from Broome).

3, 3a, 3b, 3c. *Amoria grayi* LUDBROOK, 1953
Slender and glossy smooth, without shoulders. Spire of medium height, rather straight-sided, sharply pointed. With 4 apical whorls and 3 adult whorls. Columella with 4 strong plaits. Colour variable, typically uniform cream, ash or beige, sometimes pink near the lip, and with a light brown sutural band; spirally banded and axially lined forms also occur. Columella white, deep interior brown.
10 cm. Moderately common. W.A. from Geographe Bay to the North Kimberley. This is probably the most common and most widely distributed species in the *Amoria* group. It is also quite variable in colour and markings. In northern shells the body whorl is usually uniformly coloured and lacking in markings (e.g. (3a) from Broome) but sometimes there are 2 faint brown spiral lines (e.g. (3) from Onslow). In shells from Fremantle and Geographe Bay are 2 broad brown spiral bands. In some northern specimens (e.g. (3b, 3c) from Onslow) and in deep water specimens from off Fremantle (50-120 fathoms) the bands consist of crowded brown axial dashes. Because of this variability the species needs more taxonomic study.

4, 4a. *Amoria jamrachi* GRAY, 1864
Slender, without shoulders. Protoconch rather tumid, cream-grey, with 4 apical whorls and only 2 adult whorls. Sutures channelled, lip thin, sharp and deeply notched at the suture. Adult whorls cream, marked with numerous almost straight but rather ragged brown axial lines, lines continuous and separated or coalesced.

Often with 2 central bands of axially elongate blotches which lie on the axial lines. Adult whorls with a row of ragged dark brown pre-sutural spots. Columella with 4 plaits, white, inner margin of the outer lip white, deep interior brown.
7 cm. Uncommon. Northern W.A. between North-West Cape and the North Kimberley. Compare this species with *A. turneri* page 131 (5).

5. *Amoria dampieria* WEAVER, 1960
Small and solid, with a thick lip, 3 cream-grey apical whorls in the protoconch and 2½ adult whorls. Columella with 4 strong plaits. Adult whorls creamy-white with brown almost straight axial lines. Pre-sutural spots lacking. Columella and inner margin of the outer lip white, deep interior brown.
3.2 cm. Uncommon. Monte Bello Is., W.A. to Melville Is., N.T. The shell might easily be mistaken for the eastern *Z. zebra*, but details of the protoconch show that it is not closely related to that species. Though known for some years this species was not validly described until 1960.

6, 6a. *Amoria macandrewi* SOWERBY, 1887
Rather solid with broad shoulders and a low spire. Apical whorls white or pale pink, adult whorls glossy cream or buff and characteristically marked with widely spaced (sometimes very widely spaced) zigzag dark brown lines resembling those of *A. undulata*. Usually with brown pre-sutural spots, and in some specimens (e.g. 6) 2 faint brownish spiral bands. Inner margin of the lip and the columella white, deep interior brown or purple-brown.
9 cm. Common. Intertidal and shallow water around Barrow and Monte Bello Is., W.A. This was considered to be an extremely rare species until a few years ago when the habitat was discovered at Barrow Is.

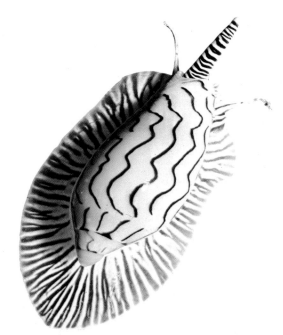

Plate 84. *Amoria macandrewi* SOWERBY, 1887, Barrow Island, W.A.   (*Photo*—Barry Wilson).

7. *Amoria ellioti* SOWERBY, 1864
Rather solid, with moderately broad shoulders and a spire of medium height. Exterior glossy cream with straight or slightly undulating axial lines of dark brown. Pre-sutural spots weak or lacking. Inner margin of lip white, deep interior coffee brown.
11 cm. Moderately uncommon. Intertidal and shallow water in the vicinity of Port Hedland and Nickol Bay, W.A. This is another species with very restricted range. *A. ellioti* seems to be closely related to *A. macandrewi,* but the axial lines are almost straight, the spire is longer and the shoulders are less pronounced in *A. ellioti.* However, a few intermediate shells recently collected on the mainland between Onslow and Dampier, W.A. cast doubt upon the relationship.

Plate 85. Volutidae (⁹⁄₁₀ x natural size)

Volutidae (Cont.)

1. *Aulicina irvinae* SMITH, 1909
Large, broad and solid, with a crown of short hollow pointed spines on the angular shoulders of the adult whorls. Exterior deep pink flecked with white. With 2 spiral bands of irregular blotches (rather than elongate lines as in *A. nivosa*). Dark lines present between the sutures and the shoulders.
13 cm. Uncommon. Cape Naturaliste to Geraldton, W.A. Typical specimens of *A. irvinae* are taken from craypots at 10 to 60 fathoms between Cape Naturaliste and Fremantle. Although shells from that region are quite distinct from *A. nivosa* there is some confusion about the identity of some large orange specimens with shoulder nodules, taken in craypots off Geraldton. These specimens are difficult to determine as one species or the other. No live specimens are yet available for study.

2, 2a. *Aulicina nivosa* LAMARCK, 1804
Rather solid, with a fairly high spire, and prominently angled shoulders which are smooth or bear only low nodules. Low axial nodules present on the apical whorls, columella with 4 strong plaits. Bluish or sometimes blue-pink with scattered white spots and 2 broad spiral bands of dark brown axial lines. Long fine brown axial lines present between the sutures and shoulders. Interior and columella orange-brown.
12 cm. Common. W.A. between Fremantle and Shark Bay. In dead shells the blue colour characteristic of the shell of living specimens is lost, and the shell becomes orange. For this reason dead shells are sometimes mistaken for *A. irvinae*.

3, 3a. *Aulicina nivosa oblita* SMITH, 1909
Broader than typical *A. nivosa*, and the shoulders bear prominent, pointed, hollow spines like those of *A. sophiae*. Colour usually more green than blue, with large and indistinct white spots (3). The 2 spiral bands sometimes consist of dark blotches, axial lines absent or indistinct.
9 cm. Common. Northern W.A. from North-West Cape to the Kimberley region. Shells from the north Kimberley region are much more tumid than specimens from further south. There is also a red-brown colour form (3a). For many years this species has been confused with *Voluta norrissii* which is a different Pacific species described by Gray (1838).

4. *Aulicina sophiae* GRAY, 1846
Rather thin and swollen, with sharply angled shoulders bearing well spaced sharp spines. Prominent nodules present on the apical whorls. Exterior fawn, mottled with pale brown and white, with 2 darker brown spiral bands, and usually 4 rows of distinct brown spots or dashes, the rows lying along the posterior and anterior edges of the spiral bands. Prominent dark axial lines present between the sutures and the shoulders of the whorls.
8 cm. Moderately uncommon. N.T. and northern Qld.

5. *Amoria turneri* GRAY [IN] GRIFFITH & PIDGEON, 1834
Slender and slightly pyriform, with a short broad spire, 4 apical whorls, 2½ adult whorls, and 4 columellar plaits. Body whorl with numerous thin axial brown lines on a creamy-white background, lines converging as they reach the sutures and ending in dark brown spots on the suture line. 2 spiral bands of indistinct brown blotches present on the body whorl. Interior light brown, inner margin of lip and columellar plaits white.
7 cm. Not very common. Northern Australia from the Kimberley, W.A. to Cairns, Qld.

6, 6a. *Amoria canaliculata* McCoy, 1869
(see also colour plate of live animal, page 120, plate 77)
Small, solid, characterized by deeply channelled sutures. 3 apical whorls white, adult whorls usually white with faint, fine red axial lines and 4 spiral rows of squarish red-brown patches (6a). Some specimens pink (6). Interior and columellar plaits white.
6 cm. Trawled off central Qld.

7, 7a, 7b. *Amoria maculata* SWAINSON, 1822
Rather solid but with a thin lip, 4½ rather tumid, cream apical whorls, 2 adult whorls and 4 columellar plaits. Colour variable, usually yellow, cream or pale orange, lacking markings except for 4 spiral rows of short thick brown lines on the body whorl and dark brown spots in the sutures (7, 7a). Interior white or pale brown, columellar plaits and inner margin of lip white.
8 cm. Intertidal and shallow waters of central and northern Qld. This species was once known as *A. caroli* IREDALE, 1924. *A. volva* GMELIN, 1791 from north Qld. is like an unspotted *A. maculata*. Should they prove to be the same then *volva* will be the correct name for the species.

Plate 86. *Amoria maculata* SWAINSON, 1822. At a depth of 10 feet, off Mossman, N. Qld.　　　(*Photo*—Neville Coleman).

Plate 87. Volutidae (natural size) ⇨

Volutidae (Cont.)

1, 1a. *Aulica flavicans* GMELIN, 1791

Typical form (1) stout and solid and rather like *A. rutila*, but usually with low nodules on the shoulders and bands of irregular brown patches instead of the blood-red markings of that species. Interior and columella orange. Some specimens with angular and heavily nodulose shoulders (1a).

10 cm. Common. N.T. and northern Qld. (also New Guinea and parts of Indonesia). A variable species for which several names have been used (e.g. *quaesita* IREDALE, 1956; *kellneri* IREDALE, 1957).

2, 2a. *Aulica rutila* BRODERIP, 1826

Heavy, short and solid, with broad rounded shoulders. Apical whorls cream and weakly ridged. Adult whorls, smooth with a white background overlain by irregular blood-red markings which tend to form spiral bands. Interior and columella brown.

10 cm. Moderately uncommon. Northern Qld.

3. *Notovoluta perplicata* HEDLEY, 1902

Solid, elongate, with prominent nodules on the angular shoulders extending anteriorly as low axial ridges for nearly ½ the length of the body whorl. Apical whorls also bear elongate nodules or folds. Columella with 6 high and sharp plaits. Exterior white with wavy orange axial lines, interior white.

8 cm. The type locality is Cairns, Qld. and few specimens are known. The illustrated specimen is the holotype which is deposited in the collections of the Australian Museum, Sydney.

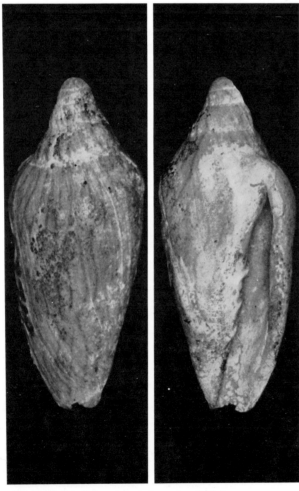

Fig. 26. *Nannamoria amicula* IREDALE, 1929

Small; spire acuminate, less than ½ the length of the aperture; anterior end also acuminate. Shoulders nodulose. Protoconch of 2½ whorls, smooth. Aperture long, narrow, outer lip thickened. Columella with 6 plaits, 3 large alternating with 3 small. Ground colour flaxen or primrose, with irregular brown axial lines and 1, 2 or 3 spiral rows of brown blotches.

3 cm. Few specimens known. N.S.W., 40 to 70 fathoms. The illustrated specimen is the holotype in the collection of the Australian Museum, Sydney.

4. *Amoria guttata* McMICHAEL, 1964

Slender, rather fusiform, with a short conical spire and a glossy smooth surface quite lacking in sculpture. Columella with 4 strong plaits, the third from the anterior end continuing as the upper edge of the anterior fasciole. Exterior cinnamon-brown with 4 darker brown spiral bands on the body whorl, overlain by irregular white marks which may take the form of axially elongated spots or short irregular white axial "stripes". Long brown axial stripes sometimes also present. Deep interior blue-white, lip edge brown.

5 cm. Uncommon. Central Qld.

5, 5a, 5b, 5c. *Zebramoria lineatiana* WEAVER & DU PONT, 1967

Very like its close relative *Z. zebra*, but distinguished by the longer and more pointed spire, and the prominent axially-elongate folds or nodules on the spire whorls. Orange, yellow, white or brown. Axial lines tend to be more crowded than in *Z. zebra*.

5 cm. Common in shallow water in central Qld. This species has been widely known as *Z. lineata* LEACH, 1814 but that name was preoccupied.

6. *Zebramoria zebra* LEACH, 1814

(see also colour plate of live animal, page 121, plate 78)

Very solid, short and ovate, without spines or nodules on the spire whorls or shoulders, surface smooth. Columella with 4 strong plaits, aperture length more than ¾ the total shell length. Exterior white, but conspicuously marked by axial lines of dark brown. Columella and lip white, interior pale orange-brown or grey.

5 cm. A common shallow water species in southern Qld. and northern N.S.W.

7, 7a. *Nannamoria parabola* GARRARD, 1960

Almost obconical, aperture long curved, and produced beyond the suture at the posterior end, equal to about ⅘ the total shell length. With 3 apical and 4 adult whorls. Shoulders of the adult whorls bear prominent spines which extend anteriorally as axial ridges. Number of columellar plaits variable (5 to 12). Pale fawn with many fine axial chestnut lines on the body whorl, and 3 spiral bands of brown blotches.

4 cm. Uncommon. Trawled in deep water off southern Qld. and northern N.S.W. *N. amicula* (Fig. 26), is like this but narrower, and does not have the prominent spines on the shoulders.

8, 8a. *Ternivoluta studeri* VON MARTENS, 1897

Rather thin, fusiform, with a broad, shallow and drawn-out anterior canal. Prominent shoulders on the whorls bear small pointed nodules. Protoconch small but bulbous, smooth. Postnuclear whorls smooth except for numerous fine spiral cords at the anterior end. Columella almost straight, columellar plaits weak, number about 6. Exterior cream or pale fawn with light brown axial lines and about 3 spiral bands of brown or cinnamon blotches on the body whorl. Lip and columella white, deep interior fawn.

6 cm. Common. Southern Qld. to northern N.S.W. Although "rediscovered" only a few years ago, this species is now commonly taken by trawlers. It is the only living member of the genus *Ternivoluta,* and appears to be the survivor of a long lineage of fossil volutes which flourished in southern Australian seas during the middle and upper Tertiary (personal communication T. A. Darragh, National Museum of Victoria). The fossil ancestral species are known under the generic name *Austrovoluta.*

9, 9a. *Relegamoria molleri* IREDALE, 1936

Rather slender but with definite shoulders and a sharply pointed spire. Columellar plaits irregular, thick and sometimes calloused. Characteristic white longitudinal ridge present just above the lip inside the aperture. Apical whorls brown and glossy. Post-nuclear whorls glossy and usually uniformly orange-pink but sometimes with very faint thin wavy red lines or 2 spiral bands of brown axial bars. A cream line runs around the sutures. Interior pink.

10 cm. Moderately uncommon. Trawled in deep water off southern Qld. and N.S.W.

10. *Ericusa sericata* THORNLEY, 1951

Rather thin for its size, smooth and glossy, without shoulders. Characterized by a depression in the middle of the body whorl near the lip, and a curiously offset protoconch. Columella with 4 plaits. Apical whorls white, early adult whorls salmon or orange-pink with few markings, body whorl salmon marked by diffuse orange-brown lines tending to demarcate triangular cream patches. A pale band runs around the centre of the body whorl. Interior cream.

13 cm. Moderately common. Trawled in deep water off southern Qld. and N.S.W.

Plate 88. Volutidae (⁹⁄₁₀ × natural size)

Volutidae (Cont.)

The genus *Volutoconus*

Of all the Australian volutes, few excite such interest as the 4 species of the genus *Volutoconus*. They are all very beautiful shells and 3 of them still rank among our most difficult species to obtain.

The first species of the genus to become known was *V. coniformis*, described by Cox in 1871 from a unique specimen found at Nickol Bay in the north of W.A. The following year Angas described *V. hargreavesi,* but was unable to give a locality for his specimen. It is a strange fact that no other specimens of either species were noted in the literature for nearly 90 years, testimony perhaps to the efficiency, or good fortune, of the early collectors.

Although the Western Australian locality for *V. coniformis* seemed well established in spite of the lack of other specimens, an Australian locality for *V. hargreavesi* was a matter of speculation. Abbott (1958) suggested that the species might come from Qld. or New Caledonia.

One of us collected a living specimen of *V. hargreavesi* on an intertidal sand bank near Port Samson, W.A., in 1956 but failed to identify it, and the specimen was subsequently stolen from a museum showcase. McMichael (1960) was first to record the discovery of new specimens of *V. coniformis* when he reported 2 shells collected at Broome, W.A. In the same paper he mentioned seeing photographs of a specimen of *V. hargreavesi* from an uncertain locality in W.A. Thus, a Western Australian home for *V. hargreavesi* was established for the first time.

Since then several specimens of *V. coniformis* have been collected along the northern coast of W.A., and additional specimens of *V. hargreavesi* have been collected from the Port Samson, Dampier Archipelago and Onslow areas. Some of these have been dredged and others picked up on beaches after storms. In addition a number of specimens of a distinctive form of *V. hargreavesi* have been taken by lobster and prawn fishermen off the central west coast of W.A. This form has been named *V. hargreavesi daisyae* WEAVER, 1967.

Much better known was *V. bednalli* BRAZIER, 1879. Although always considered to be a very rare volute, specimens turned up from time to time and a northern Australian range for the species was well established. Specimens have been collected more frequently in recent years and Abbott (1958) described the gross anatomy, radula and habitat of the species for the first time.

He also described the radula of the Qld. species *V. grossi* IREDALE, 1927 (Fig. 28, page 136) and because of the similarity of the radula and the protoconch of these 2 species he suggested that *bednalli* should be placed with *grossi* in the genus *Volutoconus*.

At the time *V. grossi* was itself considered to be a very rare shell but since then prawn trawlermen have taken large numbers of specimens from several localities off the coasts of Qld. and N.S.W. One isolated population in North Qld. is quite distinctive and has been recognized as a subspecies named *V. grossi mcmichaeli* HABE & KOSUGE, 1966.

Allan (1956) and Abbott (1958) remarked on the presence of a sharp, spine-like projection at the tip of the spire of *V. grossi*. Abbott (1958) also described a small spur at the tip of the spire of *V. bednalli*.

McMichael (1966) later pointed out a similar projection present in *V. coniformis* and *V. hargreavesi* and he proposed to call it a "calcarella", a structure also found in certain Atlantic volutes. The presence of a calcarella in all 4 species (see Fig. 27) is part of the evidence for the generic relationship between them.

Thus, in the last decade or so, substantial advances have been made in our knowledge of the genus *Volutoconus*. The Australian distributions of 2 of the species, both ranking among the world's rarest volutes, have been confirmed and extended, knowledge of the habitats and variations of all 4 species has been gained, and examination of anatomical and shell characters has indicated the generic relationship of the 4 species. An important piece of information still needed to confirm this relationship concerns the anatomy of the type species, *V. coniformis*, which remains unknown. Examination of the radula of this species, when a preserved specimen becomes available for study, should round off the story of this remarkable group of Australian volutes.

Selected references:

Because of the special interest of this genus a list of some recent references may be useful:

ABBOT, R. T. (1958). Notes on the anatomy of the Australian volutes, *bednalli* and *grossi*. J. malac. Soc. Aust., **1** (2) : 2-7, pl. 1, text figs 1, 2.

ALLAN, J. (1956). The reappearance of a rare Australian volute shell, *Amoria grossi* (IREDALE) (Mollusca, Gastropoda), a new record for New South Wales. Proc. R. Zool. Soc., N.S.W., 1954-55 : 48-49, fig. 1.

HABE, T. and KOSUGE, S.(1966). New genera and species of the tropical and sub-tropical Pacific molluscs. Venus, **24** : 312-341, pl. 19.

McMICHAEL, D. F. (1960). Notes on some Australian Volutidae. J. malac. Soc. Aust. **1** (4) : 4-13, pl. 1, text fig. 1.

McMICHAEL, D. F. (1966). A new subspecies of *Volutoconus grossi* (IREDALE) (Gastropoda : Volutidae) from North Queensland. J. malac. Soc. Aust., **1** (9) : 53-55, colour plate 14.

WEAVER, C. S. (1967). A new subspecies of *Volutoconus hargreavesi* (ANGAS, 1872) from central Western Australia (Gastropoda : Volutidae). Veliger, **9** (3) : 301-304, pl. 41.

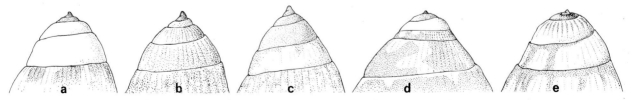

Fig. 27. The spires of the 4 species and 1 of the subspecies of the genus *Volutoconus* showing the sculptural characteristics and the terminal spine-like projection known as the calcarella.
(a) *V. hargreavesi hargreavesi*   (b) *V. h. daisyae*   (c) *V. grossi grossi*   (d) *V. coniformis*   (e) *V. bednalli*.

Plate 89. *Volutoconus grossi grossi* IREDALE, 1927.   (*Photo*—Don Byrne).

Volutidae, *Volutoconus* (Cont.)

1, 1a. *Volutoconus grossi grossi* IREDALE, 1927
See colour plate of live animal, page 135, plate 89.
Solid, elongate, slightly fusiform. Spire high, a little less than ⅓ of the total shell length. Apex with a sharp spike-like calcarella, nuclear and post-nuclear whorls smooth or weakly axially ribbed, body whorl smooth except for numerous fine axial striae, usually without prominent shoulders. Columella with 4 strong and high plaits. Anterior fasciole low. Exterior ground colour light orange-red or orange-yellow with 4 darker interrupted spiral bands of red-brown and numerous triangular white flecks and patches. Interior white.
　　17 cm. Moderately uncommon. Trawled at depths between 40 and 110 fathoms. Keppel Bay, Qld. to Port Macquarie, N.S.W. A number of distinctive forms are now known from apparently isolated populations. One of these has been recognized as a distinct subspecies (see below). The figured specimen is from Tin Can Bay, Qld. McMichael (1960) described the animal of *V. grossi* as follows: "The foot is banded with orange-red or pinkish bands on a cream background. The siphon is elongate with long equal appendages. The head is compressed, with slender tentacles, the eyes prominent at their bases."

2. *Volutoconus grossi mcmichaeli* HABE & KOSUGE, 1966
Even more slender than the typical form, axial ribs more prominent, especially on the early post-nuclear whorls and on the shoulders of the body whorl. Spiral bands prominent and black.
　　8 cm. Uncommon. Off Townsville, Qld. This northern subspecies was also named *V. grossi helenae* MCMICHAEL, 1966, but Habe and Kosuge published their description a short time earlier and their name therefore has priority.

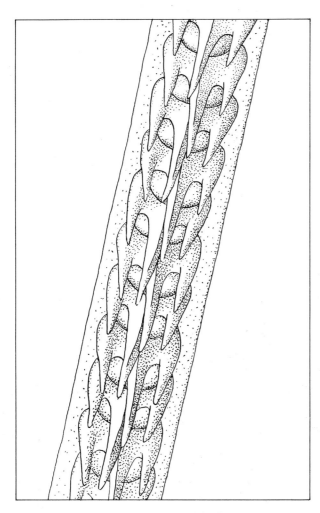

Fig. 28. Radula, enlarged section (x 100) of *Volutoconus grossi* IREDALE, 1927. Redrawn from Abbott (1958).

3. *Volutoconus bednalli* BRAZIER, 1879
Spire rather high, apical whorls with fine crowded axial striae, calcarella small and spur-like. Penultimate and body whorls with broad low axial folds at the shoulders and numerous axial growth striae. Anterior fasciole well developed, anterior canal up-turned. Columellar plaits strong and stand almost at right angles to the columella. Cream with 3 thick chocolate brown spiral lines on the body whorl (1 on spire whorls) and another at the sutures, all connected by bowed axial lines. Axial lines sometimes irregular and the spiral lines indistinct. Calcarella and apical whorls chocolate brown.
　　16 cm. Uncommon. Found on muddy intertidal sand flats and at depths to at least 50 fathoms. North Kimberley, W.A. to Torres Strait, Qld. Also found on the south coast of New Guinea and the islands of south-eastern Indonesia. The illustrated specimen is from Thursday Is.

4, 4a. *Volutoconus coniformis* COX, 1871
Solid, stout, ovate-cylindrical or fusiform. Spire low or of medium height with a small calcarella and weak axial ribs on the apical whorls. Columella with 4 strong plaits which tend to stand out nearly at right angles. Exterior looks smooth, but minutely sculptured with fine, rather straight axial striae crossed by fine, wavy, spiral striae. Very low and "broad" axial folds present on the body whorl, and a shallow spiral depression a short distance anterior to the suture. Nuclear and early post-nuclear whorls cream, penultimate and body whorls with cream ground colour and a superimposed pattern of orange-red or chestnut brown patches. Pattern consists primarily of 4 interrupted spiral bands which may coalesce, and a series of wide oblique-axial zigzag lines. Markings sometimes so crowded that the shells appear orange-red or brown, with cream "tent" marks.
　　8 cm. Very few specimens exist in collections and most known to us are worn beach shells. The known range so far is Dampier Archipelago to Cape Leveque (north of Broome) W.A. The brown shell represented by (4) is a freshly dead specimen from a beach near Port Hedland and probably represents the colour of the shell in life. (4a) is another Port Hedland beach specimen, but this one is beach worn and its orange-red colour may be faded and not the true colour of the shell in living specimens. These Port Hedland shells are typical of specimens from that area in shape. Some shells from the Broome area are more fusiform and have a higher spire. However, more specimens are needed before we can conclude that this geographic variation is consistent.

5. *Volutoconus hargreavesi hargreavesi* ANGAS, 1872
Stout and solid, spire moderately low (about ¼ of the total shell length). Apical whorls smooth or with fine axial striae, calcarella small. Penultimate and body whorls slightly shouldered, shoulders smooth, sculptured with fine axial growth striae, or sometimes with a few weak axial ribs. Anterior fasciole low and grooved. Columella with 4 very strong plaits, and sometimes an additional weak anterior one, all of which stand almost at right angles to the columella. Exterior orange-red or golden brown with 2 darker spiral bands on the body whorl and large triangular whitish patches. Interior white.
　　12 cm. Few specimens known. Sand from low tide level to at least 25 fathoms. North-West Cape to Bezout Is. (Dampier Archipelago), W.A. The shell resembles *V. grossi*, but is more squat and the calcarella is much shorter. Specimens from the southern end of the range (North-West Cape to Onslow) often have weak axial ribs (see remarks on the following subspecies). The figured specimen was collected on a beach near Port Samson after a severe storm.

6. *Volutoconus hargreavesi daisyae* WEAVER, 1967
More elongate and higher spired than the typical form of the species. Numerous and prominent axial ribs present on the nuclear and post-nuclear whorls, including the body whorl. Columella with 4 strong plaits, and 1 or 2 additional weak plaits anteriorly, i.e. more than in *V. h. hargreavesi*.
　　12.2 cm. Few specimens known, all from deep water (20 to 118 fathoms) and taken in craypots or trawls. Wedge Is. to North-West Cape, W.A. In the region between North-West Cape and Onslow there seems to be some intergradation between this west coast subspecies and the typical form from further north-east, and subspecific separation may eventually turn out to be unwarranted. The illustrated specimen was trawled at 118 fathoms at Carnarvon.

Plate 90. Volutidae, genus *Volutoconus* (1¼₀ × natural size)

Volutidae (Cont.)

BALER SHELLS—the genus *Melo*

The genus *Melo* is one of the most spectacular groups of Australian volutes. The shells are popularly known as Melons or Balers, and for sheer size they have few equals among living gastropods. The shells of some species may reach a length of more than 40 cm., and a living specimen of such size would weigh many kilograms. Balers are colourful shells with varied and attractive patterns and graceful shapes. They occupy prominent positions in most Australian collections. Balers take their name from the use to which they have been put by Australian aboriginal people. In a culture almost devoid of pottery it is natural that a ready-made pot, bowl or water container like a baler shell should have attained wide use as a domestic implement. One common use is as a container for baling water from canoes, a function to which the shape of the shell is admirably suited. Baler shells were prized trade items and specimens were carried hundreds of miles inland.

It is a curious fact that 1 or more species of baler is found on every part of the Australian coastline except in N.S.W., eastern Vic. and Tas. Abbottsmith (1969) has pointed out that in those areas the "niche" of the genus *Melo* may be filled by another very large volute, *Livonia mammilla*. Although the latter occurs in southern Qld. adjacent to populations of 2 species of *Melo*, they are not co-inhabitants, for *L. mammilla* is confined there to deep water while the balers are found in shallow water.

Fig. 29. Egg case (a dried specimen) of *Melo amphora* SOLANDER, 1786. (⅙ natural size).   (*Photo*—Keith Gillett).

Plate 91. *Melo amphora* SOLANDER, 1786. At a depth of 10 feet, Langford Reef, Qld.   (*Photo*—Neville Coleman).

The species of *Melo* are extremely variable in shell characters, especially *M. amphora*. Some hybridization may be taking place between some species where they occur together, for instance between *M. amphora* and *M. miltonis* in the Shark Bay area of W.A. These taxonomic problems concerning *Melo* have not yet been settled and much work needs to be done before the limits and relationships of the species may be satisfactorily determined.

Because of their very great size it has not been possible to illustrate fully adult specimens of all the species.

1, 1a. *Melo umbilicatus* SOWERBY, 1826
Very inflated or almost globular, lip sometimes widely flared at the posterior end. Spire low or even sunken, shoulder spines curve sharply over and almost cover the spire..Brown or cream ground colour with 2 spiral bands of darker brown blotches, and sometimes zigzag axial lines or irregular tent markings.
     25 cm. Uncommon. N.T. to southern Qld. (also unconfirmed reports from northern W.A.). (1) illustrates a young specimen beginning to develop the swollen form typical of adult *M. umbilicatus*. (1a) illustrates a young specimen more like *M. amphora* in form, but the spire is depressed.

2. *Melo georginae* GRAY [IN] GRIFFITH & PIDGEON, 1834
Ovate with an elevated spire, shoulder spines straight or slightly incurved or obsolete. Spines continue to the edge of the lip in mature specimens. Exterior orange-brown, but otherwise patterned like *M. amphora* with 2 spiral bands, axial zigzag lines, and pale triangular zones. Pattern sometimes obsolete (e.g., illustrated specimen).
     Once thought to be a small form (rarely more than 25 cm. in length) confined to the Moreton Bay area of southern Qld., but Abbottsmith (1969) reports shells as large as 29 cm from as far north as Bowen.

3. *Melo miltonis* GRAY [IN] GRIFFITH & PIDGEON, 1834
The "Southern Baler". Rather cylindrical and more elongate than other balers. Spire elevated, shoulder spines tend to curve inwards and continue to the edge of the lip even in mature specimens. Exterior purple-brown or red-brown with zigzag axial lines enclosing creamy-white triangular patches, and 2 or 3 purplish spiral bands. Interior glossy, creamy-white or yellow.
     Maxium recorded size 45 cm but shells more than 30 cm are rare. Uncommon. Western Vic. and S.A. to Geraldton, W.A.

4, 4a, 4b, 4c. *Melo amphora* SOLANDER, 1786
(see also colour plate of live animal, page 121, plate 78)

Inflated, spire low but not depressed. Shoulder spines tend to be straight or curved outwards, and do not continue to the edge of the lip in adult specimens. In some specimens spines lacking. Colour and patterns extremely variable, most commonly yellow-brown, with zigzag axial lines enclosing pale yellow triangular patches, and 2 broad pale spiral bands of darker brown blotches, bands sometimes continuous. Axial lines crowded, widely spaced, or almost lacking.
     50 cm. Fairly common. Shark Bay, W.A. to southern Qld. Often known as *M. flammeum* RÖDING, 1798 or *M. diadema* LAMARCK, 1811. In Shark Bay some specimens resemble *M. miltonis* in shape, and are difficult to determine as one species or the other. They may be hybrids. The variation in shape, colour and pattern of *M. amphora* is quite confusing. Of this species Abbottsmith (1969) remarks: "There is a huge variety of patterns and shapes, yet an unusual form is seldom confined to a single area. Similar forms do occur at vast distances apart (many thousands of miles) with no intermediate trace." The illustrated specimens come from the following localities: (4) Beagle Bay, W.A.; (4a) Yampi Sound, W.A.; (4b) Swain Reefs, Qld.; (4c) N.T.

Owing to space limitations on page 123, plate 80, this form of this species was included with shells of the genus *Melo* (opposite page).

5. *Mesericusa sowerbyi* KIENER, 1839
A pale form trawled off Ulladulla, N.S.W., differing from typical *M. sowerbyi* by the glossy pale pink shell, the fawn central band and the white pre-sutural band. Cotton (1961) described a similar shell trawled from deep water off S.A. and named it *M. stokesi*. That shell differs from the one illustrated here by its much more tumid body whorl and bulbous protoconch. The relationship of both forms remains problematical. As a provisional step we refer the specimen illustrated here to *M. sowerbyi* and suggest that *M. stokesi* may also eventually be synonymized with that species.

Plate 92. Volutidae (⁸/₁₀ × natural size)

Plate 93. *Conus anemone* LAMARCK, 1810. Cape Naturaliste, W.A. Note the gastropod *Hipponyx* attached to the spire of the cone. *(Photo*—Barry Wilson).

cones

(FAMILY CONIDAE)

THE CONIDAE IS ANOTHER FAVOURITE FAMILY of shell collectors. The shells are often brightly coloured, although they are usually covered by a thick periostracum which must be removed before the patterns are fully visible. There are more than 500 living species, most of them in the tropics. Northern Australia, particularly Qld., has a large number of species, many having wide distributions throughout the tropical Indo-West Pacific region. More than a dozen species inhabit the temperate waters of the south, although few are as colourful as their northern cousins.

The major characteristics of cone shells are the more-or-less straight-sided inverted cone (obconical) outline, the low pointed spire and the long narrow and straight aperture. They have no plaits on the columella like the olives, mitres and volutes, which some species otherwise resemble. In some species there is a tiny chitinous operculum but this is absent in others.

The generic classification of the family is still very unsatisfactory and is based mainly on shell shape and sculpture. For this reason many authors have returned to the use of the generic name *Conus* for all the species until biological studies provide a more realistic generic classification than the present one. This is the procedure that we have adopted.

Members of the family display many variations on the basic cone shape. Because of the great number of species, the general similarity of many of them, and the extreme variability of some species, cones are often very difficult to identify. The shape and colour patterns of some widespread species groups vary greatly from one locality to another and often the different locality forms have been given distinctive names.

Examples of such difficult species complexes are *Conus anemone*, *Conus catus*, and *Conus textile* and other textile cones.

The breeding habits of cones also vary. All of them lay flask-shaped capsules (see Plates 4, 95, Fig. 31), usually in groups on the underside of stones or corals. Often several females may lay their capsules in the same clutch and spawning groups of several cones under one stone are not uncommon. Some species, e.g. *Conus figulinus*, lay an egg-mass of many capsules anchored in the sand. In some cones there is a planktonic larval stage, but in others larval development takes place entirely within the capsule and development is said to be direct. At least 2 of the southern Australian cones, i.e. *C. papilliferus* and *C. anemone*, are known to have direct development. In these species, a large number of embryos develop in each capsule. The juveniles break out through an "escape hatch" at the top of each capsule and crawl off as miniature snails equipped with a tiny fragile shell already recognizable as that of a cone (Fig. 33).

Cones are predaceous animals. Most feed on worms, many on other molluscs and a few kill and eat small fish. All cones have a venom apparatus which they use to capture their prey. The venom is produced in a long coiled secretory duct which opens into the proboscis and has a muscular venom storage sac or bulb at its inner end (Fig. 30a). The radula is greatly modified from the usual gastropod type. It consists of a series of single hollow barbed shafts, like tiny harpoons (Fig. 30b) which are made in another sac, the radular sheath. A bundle of about 20 shafts may occupy the radular sac at any one time. As required, 1 of these shafts is pushed into the

140

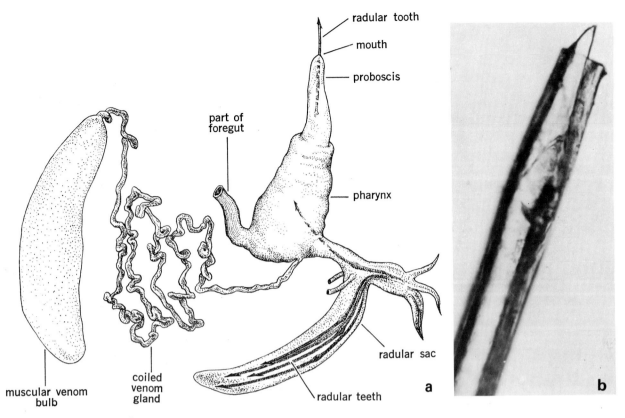

Fig. 30. (a) Venom apparatus of a typical cone. (b) Radular tooth of *Conus textile* LINNAEUS, 1758. (Photomicrograph x125).
(*Photo*—Keith Gillett).

Plate 94. *Conus geographus* LINNAEUS, 1758. This unusual photograph shows how the proboscis of the cone can be enormously expanded to devour its victim; in this case a small fish (Singapore). (*Photo*—Jack Fisher).

cavity of the proboscis where it is bathed in the venom secreted in the venom duct. The cavity of the hollow shaft fills with venom. The proboscis itself is very elastic, and can be extended out of the mouth for a distance about equal to the length of the shell. It can "feel" around in search of the victim. When the tip of the probosic touches the victim the shaft is thrust into its body and embeds itself there. The blunt end of the shaft remains held in the tip of the proboscis so that the prey is harpooned. Retraction of the proboscis then pulls it inward. The venom, a powerful neurotoxin, subdues the prey before it is swallowed whole. In some of the fish-eating species the prey may be as much as half the size of the cone and the extent to which the mouth, and indeed the whole body of the cone, is able to expand in order to swallow the victim is quite amazing (Plate 94).

DANGEROUS CONES

Many species of cones have been known to sting humans with their venom apparatus when being handled, sometimes with very mild effects. But there have been many cases of serious injury from cone stings and, at last count 16 known fatalities (Dr Alan Kohn, personal communication). Of these fatal stings, 12 are known to have been inflicted by *Conus geographus*, 2 seem to be reliably attributable to *C. textile*, and 2 were inflicted by unidentified cones. In addition severe but non-fatal injuries have resulted from the stings of *C. geographus*, *C. textile*, *C. tulipa* and *C. omaria*. Mild reactions are recorded from the stings of *C. obscurus*, *C. catus*, *C. imperialis*, *C. lividus*, *C. quercinus*, *C. sponsalis*, *C. pulicarius*, *C. litteratus* (Kohn, 1963) and *C. anemone* (G. W. Kendrick, personal communication).

From these records it appears that *Conus geographus* is the most dangerous species of cone. Scientists at one laboratory have suggested that all the reports of *C. textile* having caused injury or death to humans may have been due to misidentifications, because they found that injection of the venom of this species and other mollusc-eating cones does not have serious effects on mice. On the other hand, they found that injection of the venom of the fish-eating species such as *C. geographus*, *C. striatus* and *C. catus* caused rapid death in mice. However, the matter is not satisfactorily resolved because scientists in another laboratory obtained the opposite result in similar experiments with *C. textile*. Furthermore, although it is true that some of the fatalities originally attributed to *C. textile* have since been shown to be the result of stings by *C. geographus*, there are 2 records of fatalities fairly reliably attributable to *C. textile*. It seems that further study is needed on this matter, but in the meantime *C. textile* should be regarded as potentially lethal.

PREVENTION OF CONE STINGS

All living cone shells should be regarded as dangerous and handled as little as possible. It is recommended that collectors push a living specimen over with a knife or stick so that the animal retracts into the shell before it is picked up. A specimen should be picked up by the back of the shell near the spire (far away from the anterior end where the proboscis is located) and put quickly into a thick collecting bag or pail. On no account should living cones be held in the palm of the hand or put into a pocket.

Plate 95. *Conus papilliferus* Sowerby, 1834, with egg capsules. At a depth of 15 feet, Port Jackson, N.S.W. (*Photo*—Neville Coleman).

Plate 96. *Conus gloriamaris* CHEMNITZ, 1777. Although this species has yet to be found in Australian coastal waters this unique laboratory photograph was included for it clearly shows siphon, eye stalks, proboscis and foot of the animal (the latter partially covered with sand grains). *C. gloriamaris* has always been regarded as one of the most sought after cone shells by collectors and a live specimen (length 74 mm.) was kindly forwarded to Sydney for study by Walter Gibbins, Honiara, British Solomon Islands. *(Photo*—Terry Barlow).

TREATMENT OF CONE STINGS

The proper treatment for cone stings is still uncertain, although the following statement may be helpful:

"Lancing of the wound and removal of as much venom as possible by suction and haemorrhage, and application of a tourniquet where feasible, are recommended immediate procedures. Subcutaneous injection of adrenalin may be useful for vasoconstriction or augmentation of heart beat in severe cases." (Kohn, 1963).

If stung by a cone, a person should consult a doctor as quickly as possible (no matter what species of cone is involved). Artificial respiration may also be necessary. Pain may or may not be associated with the wound, but an early symptom is tingling in the fingers which quickly spreads to other parts of the body, particularly the lips. Eyesight may also be affected, and numbness and swelling in the area of the wound are also common symptoms.

Selected references:

CERNOHORSKY, W. O. (1964). The Conidae of Fiji. Veliger **7** : 61-94, pls 12-18.

CLELAND, J. B. & SOUTHCOTT, R. V. (1965). *Injuries to man from Marine Invertebrates in the Australian Region.* Spec. Rep. Ser. Natn Hlth med. Res. Counc. Aust. no. 12 : 1-282.

COTTON, B. C. (1948). Australian cone shells. Roy. Soc. S. Aust., Conchology Club Publ. No. 5, 4 pp., 3 pls.

DAUTZENBERG, Ph. (1937). Résultats scientifiques du voyage aux Indes Orientales Néerlandaises; gastéropodes marins; famille Conidae. Mém. Brux. Inst. Royal des Sci. Nat. de Belg. **2** (18) : 1-284, col. pls 1-3.

ENDEAN, R. & RUDKIN, CLARE (1963). Studies on the venoms of some Conidae. Toxicon **1** : 49-64.

KOHN, A. J. (1959). The ecology of the genus *Conus* in Hawaii. Ecol. Monogr. **29** : 47-90.

KOHN, A. J., SAUNDERS, P. H. & WIENER, S. (1960). Preliminary studies on the venom of the marine snail *Conus.* Ann. N.Y. Acad. Sci. **90** : 706-725.

KOHN, A. J. (1963). Venomous marine snails of the genus *Conus.* pp. 83-96 in *Venomous and Poisonous Animals and Noxious Plants of the Pacific Area,* ed. H. L. Keegan & W. V. MacFarlane, Oxford, Pergamon Pr.

MARSH, J. A. & RIPPINGALE, O. H. (1964). *Cone Shells of the World.* Brisbane, Jacaranda Pr. 166 pp., 22 col. pls.

SARRAMEGNA, R. (1965). Poisonous gastropods of the Conidae family found in New Caledonia and the Indo-Pacific. Tech. Pap. S. Pacif. Commn. no. 144: I-IV, 1-24, 30 figs.

1. *Conus geographus* Linnaeus, 1758
(see also colour plate of live animal, page 141, plate 94)
Thin and cylindrical. Spire low, coronate. Body whorl smooth with convex sides and nodulose shoulders. Aperture wide, especially anteriorly. Exterior bluish or pink-white, with a faint reticulate pattern of brown lines enclosing small white tent-like areas, and several broad bands of brown blotches, interior white.
  13 cm. Uncommon. In sand under coral. Indo-West Pacific; North-West Cape, W.A. to eastern Qld. A fish-eating species with a highly toxic venom responsible for at least 12 human fatalities. Live specimens should be handled with extreme caution.

2. *Conus tulipa* Linnaeus, 1758
Thin, subcylindrical to bulbous. Spire short with concave sides, shoulders of spire whorls angulate and finely beaded, shoulder of body whorl rounded. Surface smooth except for fine raised cords at the anterior end. Aperture wide, outer lip curved. Exterior blue-white or violet, with many thin interrupted brown spiral lines, and 2 spiral bands of red-brown patches, interior violet. Periostracum moderately thin, yellow-brown, with rows of raised pustules.
  8 cm. Uncommon. Under corals. Indo-West Pacific; eastern Qld. A fish-eating species known to be dangerous to man.

3. *Conus virgo* Linnaeus, 1758
Solid, elongately obconic. Spire low with concave sides, sides of body whorl straight to concave, shoulders more or less rounded. Body whorl dull, smooth and sculptured only by weak spiral cords which are stronger anteriorly. Aperture straight and narrow. Exterior uniformly pale yellow except for a purple anterior tip, interior white. Periostracum thick, brown.
  15 cm. Common. In sand and among coral. Indo-West Pacific; eastern Qld.

4. *Conus figulinus* Linnaeus, 1758
Solid, obconic to pyriform. Spire of medium height with straight to concave sides and elevated apex, spire whorls rounded. Shoulders rounded, sides of body whorl slightly convex, smooth except for fine granulose spiral cords anteriorly. Aperture moderately wide, straight. Exterior light to dark brown with many fine spiral brown lines, interior white.
  9 cm. Moderately common. On sand. Indo-West Pacific; eastern Qld.

5. *Conus striatus* Linnaeus, 1758
Elongate, subcylindrical, rather solid. Spire of medium height, sutures channelled. Body whorl with convex sides, sharply angulate shoulders, and sculptured with numerous prominent spiral striae. Aperture moderately wide, anterior canal wide. External colour variable, usually pink-white with large irregular blotches of purple-brown which are most dense in 2 broad central spiral bands, interior white.
  10 cm. Moderately common. In sand under corals. Indo-West Pacific; Pt. Cloates, W.A. to eastern Qld. A dangerous fish-eating species.

Fig. 31. Egg capsules of *Conus striatus* Linnaeus, 1758, Qld.
(*Photo*—Keith Gillett).

6. *Conus marmoreus* Linnaeus, 1758
Solid, obconic. Spire low, coronate, with concave sides. Body whorl with angular shoulders and nearly straight sides, smooth except for weak spiral cords anteriorly. Aperture moderately narrow and straight. Exterior black with large white approximately triangular patches, deep interior pink, becoming white near the lip.
  10 cm. Common. On reefs. Indo-West Pacific; eastern Qld. Albinistic colour forms occur at some localities. Believed to be a dangerous species.

7. *Conus imperialis* Linnaeus, 1758
Solid, obconic. Spire low to nearly flat, heavily coronate. Body whorl with angulate, nodulose shoulders and slightly concave sides, sculptured at the anterior end with weakly nodulose spiral cords which become obsolete posteriorly. Aperture narrow and straight. Exterior white or cream with 2 broad central brown spiral bands, numerous brown spiral lines interrupted by white spots, and spiral rows of fine brown dashes and spots, anterior blue-grey, interior white, brown at the anterior end.
  11 cm. Common. In sand and among coral rubble. Indo-West Pacific; eastern Qld.

8. *Conus leopardus* Röding, 1798
Solid, obconic, spire low. Body whorl smooth but not polished, with sharply angulate shoulders and almost straight sides. Aperture narrow and straight. Body whorl pale grey or pale yellow, with many spiral rows of small irregular dark grey or dark brown spots, irregular crowded dark brown wavy lines radiate from the apex across the spire whorls, interior white. With a thick, matted yellow periostracum.
  20 cm. Moderately common. In sand and among corals. Indo-West Pacific; eastern Qld. Resembles *C. litteratus* but lacks the polish, the yellow spiral bands, and the violet-brown coloration at the anterior end characteristic of that species.

9. *Conus litteratus* Linnaeus, 1758
Solid, obconic, spire flat, sometimes slightly concave. Body whorl smooth and polished, with sharply angulate shoulders and straight sides. Aperture narrow and straight. Exterior white with 3 indistinct yellow spiral bands, and many spiral rows of large squarish or rectangular black to chocolate brown spots, wavy black or chocolate brown lines radiate from the apex across the spire whorls; interior white. With a thick matted yellow periostracum.
  13 cm. Common. In sand and among corals. Indo-West Pacific; eastern Qld. Compare with *C. leopardus*.

10. *Conus vexillum* Gmelin, 1791
Solid, obconic, spire low. Spire whorls spirally striate, body whorl smooth, sculptured only by weak irregular growth striae. Aperture wide and straight, outer lip thin and fragile. Exterior rich orange-brown with 2 poorly defined white spiral bands, 1 at the shoulder and 1 central, anterior end dark brown, large dark brown blotches present on the spire whorls, interior white. Periostracum thick, olive-brown.
  11 cm. Moderately common. On reefs and in sand under corals. Indo-West Pacific; Pt. Cloates, W.A. to eastern Qld.

11. *Conus quercinus* Solander, 1786
Solid, heavy, obconic. Spire spirally striate, of medium height with concave sides and a prominently pointed apex. Body whorl with rounded shoulders and nearly straight sides, smooth except for numerous fine irregular spiral cords. Aperture moderately wide and straight. Exterior yellow, sometimes with crowded fine light brown spiral lines on the body whorl, interior white, anterior end of the columella often calloused and tinged with green.
  12 cm. Common. On sand. Indo-West Pacific; eastern Qld.

12. *Conus betulinus* Linnaeus, 1758
Solid, heavy, obconic. Spire flat with an elevated apex. Body whorl with rounded shoulders, smooth except for spiral cords anteriorly. Aperture wide. Exterior dark cream or yellow with spiral rows of widely spaced brown spots, brown radial streaks present on the spire, interior white to cream.
  13 cm. Moderately common in some parts of its range but not in Australia. Indo-West Pacific; eastern Qld.

Plate 97. Conidae ($\frac{8}{10}$× natural size)

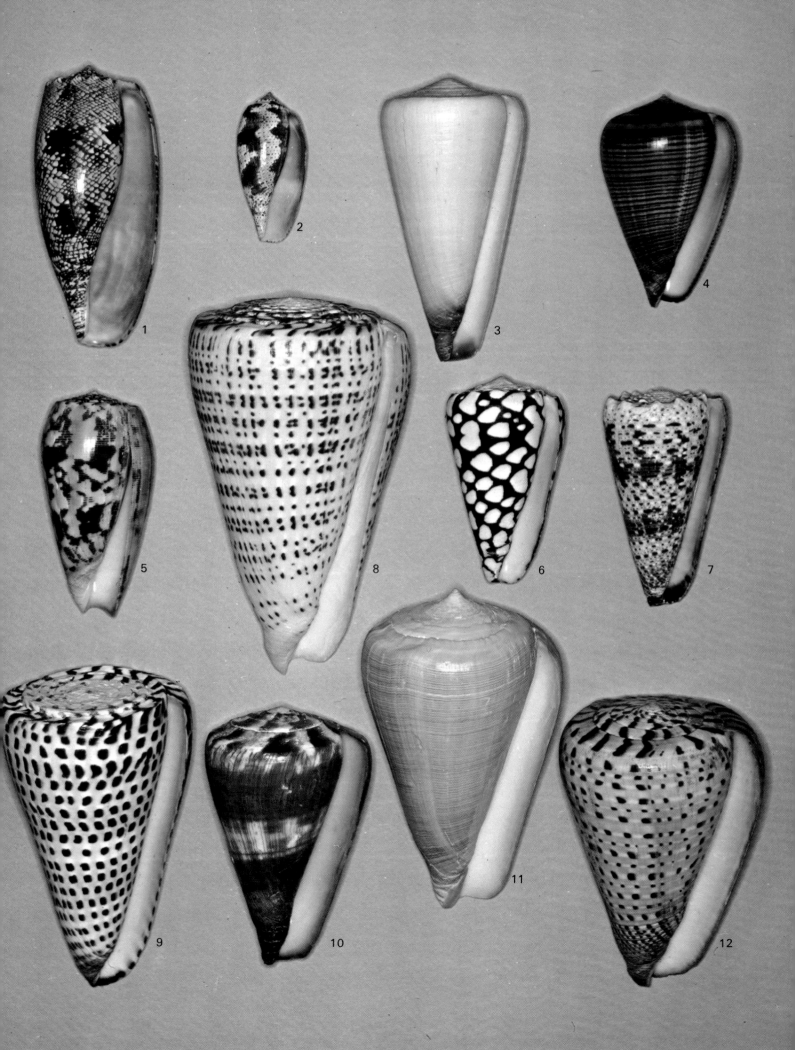

Conidae (Cont.)

1. *Conus episcopus* HWASS [IN] BRUGUIÈRE, 1792

Elongate, cylindrical. Spire moderately low with concave sides, post-nuclear whorls broad and rounded but the apex itself a tiny pointed tip. Shoulders of the body whorl slightly angulate, body whorl with convex sides, sculptured with fine spiral cords. Exterior rich red-brown with spiral rows of tiny white spots and large white tent-like areas, interior white.

    7 cm. Uncommon. In sand under coral slabs. Indo-West Pacific; eastern Qld. The broad early post-nuclear whorls distinguish this species from *C. omaria* and *C. aulicus* which it resembles.

2, 2a. *Conus omaria* HWASS [IN] BRUGUIÈRE, 1792

Slender, cylindrical. Spire rather low with a minute apex. Shoulders moderately angulate, surface smooth and glossy, sculptured only by very fine spiral cords which are stronger at the anterior end. Anterior canal broad. Exterior brown with numerous fine interrupted white spiral lines, and small tent-like areas of white which tend to concentrate in 3 spiral zones, interior white.

    7 cm. Uncommon. In sand under stones and corals. Indo-West Pacific; eastern Qld. (W.A.-?). Another variable species of the "tent-cone" group and should be regarded as possibly dangerous.

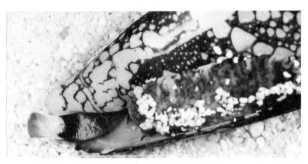

Fig. 32. Close-up study of *Conus omaria* HWASS [IN] BRUGUIÈRE, 1792, Qld.   *(Photo*—Keith Gillett).

3, 3a. *Conus textile* LINNAEUS, 1758

(see also colour plate of live animal, page 13, plate 4)

Spire moderately high with slightly concave sides. Full grown specimens rather bulbous with convex sides, but this is less marked in younger specimens. Whorls smooth and glossy with very fine spiral cords which become stronger toward the anterior end. Aperture widening anteriorly. Exterior orange-brown with patches of darker brown containing undulating axial lines and a crowded pattern of tent-like white areas, interior white.

    10 cm. Common. In sand under stones and corals. Indo-West Pacific; Pt. Cloates, W.A. to northern N.S.W. This is the best known of the "tent-cones", and its sting is said to be responsible for at least two human fatalities.

4. *Conus aulicus* LINNAEUS, 1758

Elongate and subcylindrical with a pointed rather straight-sided spire of medium height. Body whorl with rounded shoulders and convex sides, sculptured with very fine spiral cords. Aperture curved, widening anteriorly. Exterior rich red-brown, with sharply outlined tent-like areas of pink-white, deep interior yellow or orange, becoming white toward the lip.

    14 cm. Uncommon. In sand, under stones and corals. Indo-West Pacific; eastern Qld. There are reports of this species stinging humans but not fatally.

5, 5a, 5b, 5c, 5d. *Conus victoriae* REEVE, 1843

Spire moderately high with slightly concave sides and a pointed apex. Shoulders weakly angulate or rounded, sides of body whorl convex, surface smooth except for weak spiral cords anteriorly. Aperture widening anteriorly. External colour variable, the most common form (5b) with a dense pattern of golden brown reticulate lines enclosing tiny tent-like areas, overlain by blue-tinted patches and bands of darker brown, interior white or blue-white.

    7 cm. Common. Under and among stones. North-West Cape, W.A. to the N.T. Collectors are at present using different names for several colour forms of this species. A particularly dark form (5c) is being called *C. cholmondeleyi* MELVILL, 1900 but that is a distinct East African species. Forms lacking the blue tints (5, 5a) are being called *C. complanatus* SOWERBY, 1866 but these do not warrant taxonomic separation. There are also forms lacking the golden brown reticulations (5d) but, fortunately, these have so far escaped the attention of the "splitters".

6. *Conus nodulosus* SOWERBY, 1864

Very like *C. victoriae* but the colour pattern paler, lacks any trace of bluish tints, and tent-like areas usually larger. Interior characteristically pink. Brown bands absent or obsolete.

    5 cm. Common. Shallow reefs. Fremantle to Shark Bay, W.A. This may be only another distinctive colour form of *C. victoriae* but it occurs along a different part of the W.A. coast and within its range the coloration is consistent and distinct from that of the more northern *C. victoriae*. It may eventually be regarded as a subspecies.

7, 7a. *Conus generalis* LINNAEUS, 1767

Solid, obconic, elongate. Spire flat, but with a pointed protruding apex. Shoulders angulate, sides of the body whorl nearly straight, surface smooth except for a few weak spiral cords anteriorly. Aperture straight and narrow. External colour variable, usually orange-brown or chocolate brown with 3 interrupted bands of white, irregular radial rays present on the spire, interior white.

    9 cm. Common. In sand. Indo-West Pacific; North-West Cape, W.A. to eastern Qld.

8, 8a. *Conus ammiralis* LINNAEUS, 1758

More-or-less obconic, but the spire moderately high, with concave sides and a sharp apex. Shoulders rather angulate, sides of body whorl nearly straight, surface smooth. Colour variable in detail, but usually brown with narrow dark interrupted spiral lines, white tent-like areas of varying size, and 3 bands of yellow or pale brown with minute tent-like areas, irregular radial rays present on the spire, interior white.

    8 cm. Uncommon. Indo-West Pacific; eastern Qld. *C. temnes* IREDALE, 1930 is probably a deep water form of this intricately patterned species.

9, 9a. *Conus miles* LINNAEUS, 1758

Solid, obconic. Spire of medium height with slightly concave sides. Body whorl with more or less rounded shoulders, surface smooth except for widely spaced spiral cords anteriorly. Exterior white with thin undulating axial orange-brown lines, light orange-brown patches, and 2 broad dark brown spiral bands, 1 central and the other at the anterior end, external colour shows through to the interior.

    8 cm. Common, on rocky and coral reefs. Indo-West Pacific; North-West Cape, W.A. to eastern Qld.

10. *Conus capitaneus* LINNAEUS, 1758

Obconic, broad and angulate at the shoulders. Spire moderately low, with concave sides and strong spiral striae. Body whorl straight-sided, sculptured with weak spiral cords which become obsolete posteriorly. Aperture rather wide. Exterior yellowish to olive-brown, darker at the anterior end, with white spiral bands at the shoulder and centre of the body whorl, each containing dark brown spots, radiating brown rays present on the spire, interior violet.

    8 cm. Moderately common, under corals. Indo-West Pacific; Pt. Cloates, W.A. to eastern Qld.

11, 11a. *Conus trigonus* REEVE, 1848

Similar to *C. capitaneus* in shape, but not as broad at the shoulders, and the sides of the spire less concave. Exterior yellow or orange-brown, with interrupted spiral brown lines and 3 white spiral bands, 1 at the shoulder, 1 at the centre and 1 at the anterior end. Irregular brown radial rays present on the spire, interior white.

    7 cm. Moderately common but not often collected alive. North-West Cape, W.A. to N.T.

12. *Conus distans* HWASS [IN] BRUGUIÈRE, 1792

Solid, obconic, spire very low with concave sides, coronated. Body whorl smooth with angulated nodulose shoulders and straight to concave sides. Aperture straight, narrow. Exterior pale yellow-brown or olive-green, with 2 indistinctly defined white spiral bands, 1 at shoulder and 1 central, anterior end dark brown, spire with a spiral row of large squarish brown spots, interior pale violet. Periostracum thick, nodulose and brown.

    9 cm. Moderately uncommon. On coral reefs. Indo-West Pacific; eastern Qld.

Plate 98. Conidae ($\%_{10}$ ˣ natural size)

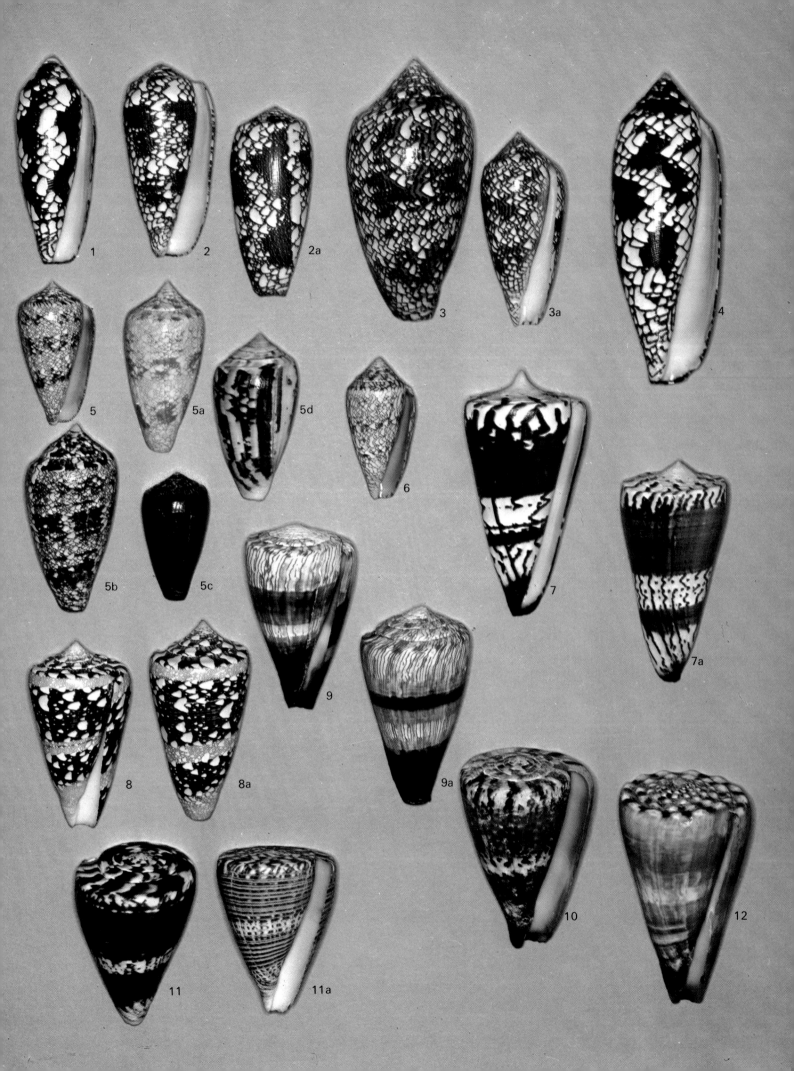

Conidae (Cont.)

1. *Conus glans* HWASS [IN] BRUGUIÈRE, 1792
Solid, slender, biconic to cylindrical. Spire of medium height, with convex sides and a sharply elevated apex. Body whorl with rounded shoulders, convex sides, sculptured with strong raised spiral cords, anterior cords beaded. Aperture narrow posteriorly, widening slightly anteriorly. Exterior dark blue with white clouding, sometimes with purple-brown spiral bands, interior purple.
 3 cm. Uncommon. Indo-West Pacific; North-West Cape, W.A. to eastern Qld.

2, 2a. *Conus clarus* SMITH, 1881
Obconic, spire low, usually with concave sides. Shoulders angulate, sides of body whorl almost straight, smooth except for spiral grooves anteriorly. Body whorl uniform white or yellow, spire similarly coloured or with faint radial streaks, interior white.
 5.5 cm. Moderately common. A sand dweller. North-West Cape to North Kimberley, W.A.

3, 3a, 3b. *Conus klemae* COTTON, 1953
Biconic, spire moderately high, pointed, with concave sides and shouldered whorls. Shoulder of body whorls angulate, surface smooth except for a few raised anterior spiral cords. Ground colour uniform reddish, orange-brown, pink or yellow, sometimes overlain by an irregular pattern of darker zigzag axial lines or maculations, always with 3 pale spiral bands, at the shoulder, centre, and anterior end. Anterior band always paler than the ground colour of the rest of the body whorl. Brown patches present on the shoulder within the posterior spiral band.
 7.5 cm. Uncommon. Under and among stones. S.A. to Geraldton, W.A.

4, 4a. *Conus dorreensis* PÉRON, 1807
Biconic, spire moderately high and prominently coronated. Body whorl with convex sides and prominent shoulder nodules, sculptured with spiral rows of pits. White or pale pink, but a thin green-yellow closely adherent periostracum gives a characteristic appearance. Black spiral periostracal bands often present on the shoulder and behind the anterior end.
 4 cm. Abundant at some localities. Intertidal reefs. Cape Leeuwin to Barrow Is., W.A. Once better known as *C. pontificalis* LAMARCK, 1810.

5, 5a. *Conus cocceus* REEVE, 1844
Spire of medium height, with strong spiral striae, convex sides, and a tiny papilliform apex. Body whorl with rounded shoulders and convex sides, surface smooth and glossy, but sculptured with minute punctate spiral striae which become fine spiral cords anteriorly. Pale rose-pink or yellow, with irregular white and orange blotches, and spiral rows of white flecks, interior white or pink. Some specimens white.
 4 cm. Uncommon. Shallow reefs. Albany to Cape Naturaliste, W.A. Said to be a fish-eater and thus it is potentially dangerous.

6, 6a, 6b, 6c. *Conus anemone* LAMARCK, 1810
(see also colour plate of live animal, page 140, plate 93)
Bulbous to biconic; spire spirally striate, short with concave sides, or high and shouldered, or straight-sided. Body whorl with weak spiral striae and weak anterior spiral cords. Colour variable, the most common colour form (6a) is blue-grey flecked with white, with a ragged reticulate pattern of axial lines and maculations, and darker blue or brown spiral bands. Orange markings (6c) in some specimens, others uniformly pink (6, 6b) yellow or white.
 8 cm. Very common. Under and among stones. Central N.S.W. to Geraldton, W.A. Many names have been applied to forms of this species. (6) an unusual bulbous pink specimen taken at 10 fathoms in Geographe Bay, W.A., resembles the type specimen of *C. peronianus* IREDALE, 1931 from N.S.W., which is probably only a form of *C. anemone*.

7. *Conus compressus* SOWERBY, 1866
Like *C. anemone* but more elongate, biconic, with a high and strongly shouldered spire.
 6 cm. Uncommon. S.A. This is probably only yet another form of *C. anemone*. Although the illustrated shell from South Australia has orange markings, shells of similar shape from southern W.A. have the typical bluish coloration of *C. anemone*.

8, 8a, 8b. *Conus novaehollandiae* A. ADAMS, 1854
Very like *C. anemone* but rather more bulbous, and the spiral striae are slightly coarser. Colour variable, typically blue-grey with white flecks and large tan to chocolate-brown blotches.
 4.5 cm. Very common. North-West Cape to Cape Leveque, W.A. Another member of the *C. anemone* complex, and perhaps would be better regarded as a subspecies of that species.

9. *Conus segravei* GATLIFF, 1891
Biconic, spire high, rather straight-sided. Sides of body whorl almost straight, shoulders angulate, sculptured with fine spiral subsutural striae, body whorl smooth except for a series of fine spiral grooves at the anterior end. Exterior cream or white, with pale pink or orange patches containing tent-like white marks, sometimes with spiral rows of darker dots, interior rose-red or orange.
 3 cm. Uncommon. A sand dweller. Vic. to Cape Leeuwin, W.A.

10. *Conus papilliferus* SOWERBY, 1834
(see also colour plate of live animal, page 142, plate 95)
Obconic, broad and angulate at the shoulders, with slightly convex sides. Spire spirally striate, low with a pointed apex. Body whorl with weak spiral cords at the anterior end. Colour variable, usually whitish with irregular blue-grey blotches, overlain by irregular brown blotches and many rows of alternately white and brown fine spots, interior bluish or brownish-white.
 2.5 cm. Moderately common. N.S.W.

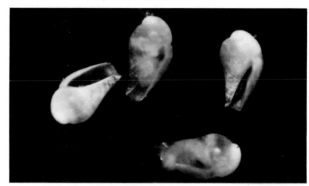

Fig. 33. *Conus papilliferus* SOWERBY, 1834, juveniles recently hatched from egg capsules (x18), N.S.W. (*Photo*—Keith Gillett).

11, 11a. *Conus aplustre* REEVE, 1843
Obconic, broad and subangulate at the shoulders, with slightly convex sides. Spire of medium height with a pointed apex and fine spiral striae. Body whorl smooth except for weak spiral cords anteriorly. Exterior base colour blue-grey, overlain by fine spiral lines of brown and white dashes or dots, and often with a central band of irregular brown patches, brown patches present on the spire, interior deep violet becoming brown towards the lip.
 2.5 cm. Moderately common. N.S.W. May be only a colour form of *C. papilliferus*.

12, 12a, 12b, 12c, 12d. *Conus rutilus* MENKE, 1843
Very small, thin, obconic. Spire low and coronated. Sides of body whorl slightly convex, surface smooth except for spiral striae at the anterior end. Colour variable, usually pink or red-brown, but sometimes yellow or violet, usually with 2 or more spiral bands of squarish darker spots on the body whorl.
 1.5 cm. Common in beach drift but rarely collected alive. N.S.W. to Fremantle, W.A. *C. smithi* ANGAS, 1877 is a strongly patterned form of this species.

13. *Conus spectrum* LINNAEUS, 1758
Thin and rather bulbous. Spire low, with fine spiral striae, concave sides and a sharp apex. Outer lip curved. Body whorl with angulate shoulders, shiny smooth but with minute spiral striae and weak spiral cords at the anterior end. Exterior ivory-white or cream, with fine wavy brown axial lines, and irregular patches of brown which tend to form 3 or 4 spiral bands and sometimes cover much of the surface, brown patches present on the spire, interior white.
 6 cm. Moderately common, in sand. Central Indo-West Pacific; Shark Bay, W.A. to eastern Qld.

14. *Conus planorbis* BORN, 1778
Spire low, with straight to concave sides and channelled sutures. Body whorl with angulate shoulders and slightly convex sides, smooth except for weak spiral cords anteriorly. Exterior light yellow-brown, with a brown anterior end, a poorly defined white spiral band at the shoulder and another at the centre, and sometimes narrow continuous or interrupted brown spiral lines, semicircular brown flecks present on the spire whorls, interior white.
 6 cm. Moderately common. Indo-West Pacific; Barrow Is., W.A. to eastern Qld. The spiral lines are uncommon in Australian specimens, but characteristic of the specimens from other localities.

Plate 99. Conidae (1 1/10 x natural size) ⇨

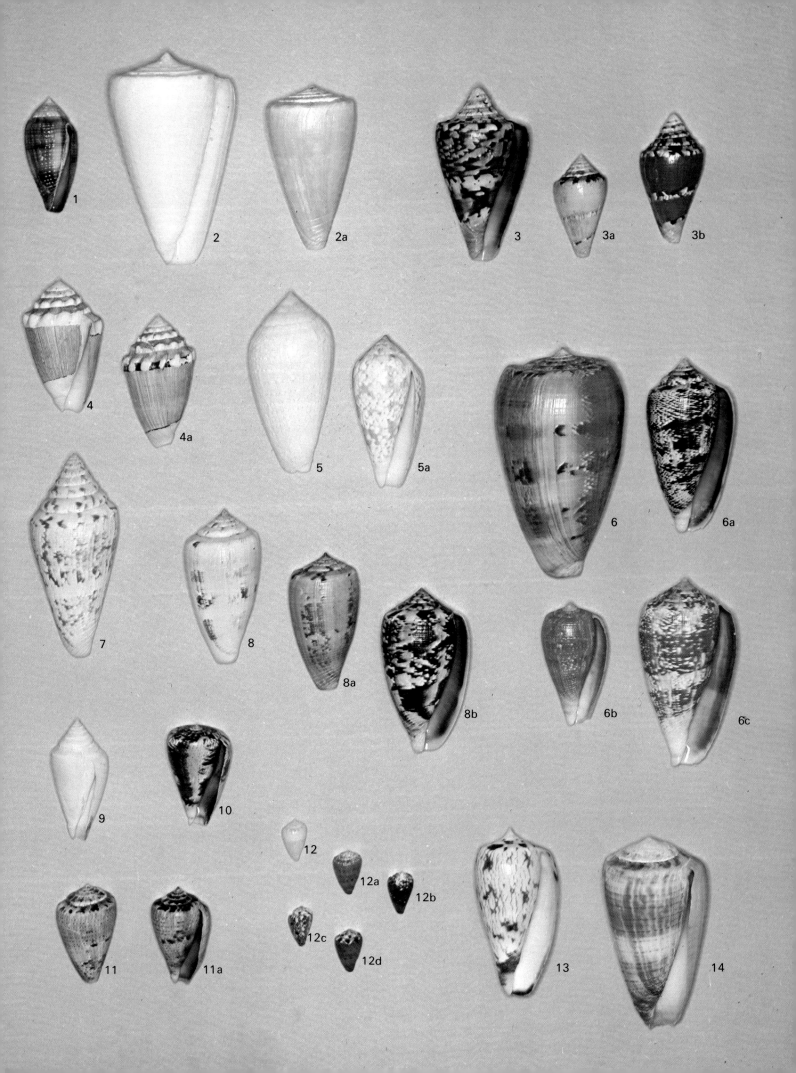

Conidae (Cont.)

1. *Conus obscurus* SOWERBY, 1833
Thin and fragile, slender, cylindrical. Spire of medium height, with an elevated pink apex, spirally striate, shoulders rounded and smooth. Body whorl smooth except for minute spiral striae and a few weak spiral cords anteriorly. Aperture wide, especially anteriorly. Exterior violet, with spiral rows of tiny white or brown spots or dashes and 3 broad spiral bands of brown blotches, interior violet.
  4 cm. Uncommon. In sand under corals. Indo-West Pacific; Barrow Is., W.A. to eastern Qld. This fish-eating species is similar and related to *C. geographus,* although much smaller. Cases of *C. obscurus* stinging humans have been reported, but none have proven fatal.

2, 2a. *Conus eburneus* HWASS [IN] BRUGUIÈRE, 1792
Solid, obconic. Spire low or of medium height, with concave sides. Strong spiral striae present on the spire. Body whorl with rounded shoulders, and almost straight sides, smooth except for weak spiral grooves anteriorly. Exterior white, with a variable number of yellow spiral bands and rows of large squarish brown spots, interior white.
  6 cm. Common. In sand. Indo-West Pacific; eastern Qld.

3. *Conus tessulatus* BORN, 1778
Solid, obconic. Spire of medium height with concave sides, deeply spirally grooved. Body whorl with rounded or angulate shoulders, nearly straight sides, smooth except for fine spiral striae and several spiral grooves anteriorly. White, with spiral rows of large rectangular orange-red spots which are most dense in 2 central spiral bands, anterior end violet, interior white.
  4.5 cm. Moderately common. In sand. Indo-West Pacific; eastern Qld.

4. *Conus ebraeus* LINNAEUS, 1758
Solid, stout, obconic. Spire of medium height with nearly straight sides. Body whorl with convex sides and more or less rounded, weakly nodulose shoulders, sculptured with spiral cords which become obsolete posteriorly. White, pink or cream, with 3 spiral bands of large rectangular blackish blotches, 1 around the centre of the body whorl, 1 below the shoulders and 1 near the anterior end, interior with brown pattern corresponding to the external banding.
  4 cm. Common. On rocky reefs. Indo-West Pacific; Houtman Abrolhos, W.A. to southern Qld.

5, 5a. *Conus chaldaeus* RÖDING, 1798
Solid, stout, slightly biconic. Spire of medium height with nearly straight sides. Body whorl with convex sides and rounded, weakly nodulose shoulders, sculptured with weakly beaded spiral cords which become obsolete posteriorly. Blackish, with white spiral lines at the shoulders and the centre of the body whorl, and irregular wavy axial white streaks, interior white, margin of lip black.
  4 cm. Common. On rocky reefs. Indo-West Pacific; Houtman Abrolhos, W.A. to southern Qld. Often confused with *C. ebraeus* but the colour differences are consistent.

6. *Conus arenatus* HWASS [IN] BRUGIÈRE, 1792
Solid, stout, rather cylindrical. Spire moderately low, coronated, with slightly convex sides and a sharp apex. Body whorl with convex sides and rounded, usually nodulose shoulders, surface sculptured with minute spiral striae which become fine cords anteriorly. Aperture rather wide. White or cream, with numerous tiny brown spots arranged to form a pattern of irregular axial and spiral bands, interior white.
  6 cm. Common. In sand. Indo-West Pacific; Pt. Cloates, W.A. to eastern Qld.

7. *Conus pulicarius* HWASS [IN] BRUGUIÈRE, 1792
Solid, obconic or tending to be cylindrical. Spire coronated, moderately low with convex or nearly straight sides. Body whorl with convex sides and rounded nodulose shoulders, surface minutely striate and with weak spiral cords anteriorly. White, with groups or spiral bands of squarish brown spots and sometimes indistinct yellow spiral bands, interior white, pale yellow or pink.
  6 cm. Common. In sand. Indo-West Pacific; Pt. Cloates, W.A. to eastern Qld. May be confused with *C. eburneus* but the spire coronations usually serve to distinguish it.

8. *Conus nussatella* LINNAEUS, 1758
Cylindrical and very narrow. Spire of medium height, with convex sides and slightly channelled sutures. Body whorl shoulders rounded, surface sculptured with fine granular spiral cords and numerous striae. White, with longitudinal and spiral rows of orange-brown blotches, dark brown spots arranged along the white spiral riblets, interior white.
  6 cm. Uncommon. Found among living "finger" corals. Indo-West Pacific; eastern Qld.

9, 9a. *Conus suturatus* REEVE, 1844
Solid, stout, obconic. Spire moderately low with concave sides and pointed apex. Body whorl with more or less rounded shoulders and convex sides, surface glossy smooth except for spiral grooves anteriorly, and deep spiral striae on the spire. White, with spiral bands of pink or rose, anterior tip mauve; interior white.
  4 cm. Moderately common. In sand. Central Indo-West Pacific; Dampier Archipelago, W.A. to eastern Qld. May be only a colour form of *C. tessulatus.*

10, 10a. *Conus scabriusculus* DILLWYN, 1817
Spire of medium height, with markedly convex sides and pointed apex. Body whorl with rounded shoulders and convex sides. Aperture narrow. Spire smooth, but behind the shoulder the body whorl sculptured with numerous prominently beaded spiral cords. External colour variable, usually white or pale violet, with large chocolate brown blotches which may be fused to form 2 broad bands, anterior tip violet, brown blotches present on the spire, interior pale violet.
  5 cm. Uncommon. Pacific; N.T. to eastern Qld.

11. *Conus flavidus* LAMARCK, 1810
Solid, elongately obconic. Spire low, with straight to convex sides, spirally striate. Body whorl with sharply angulate shoulders and straight to concave sides, smooth except for weakly-beaded spiral cords anteriorly. Aperture narrow, straight. Exterior light orange-brown to red-brown, with 2 narrow white spiral bands, 1 at shoulder and 1 at centre, anterior end dark purple, interior violet.
  6 cm. Common. On reefs. Indo-West Pacific; eastern Qld.

12. *Conus lividus* HWASS [IN] BRUGUIÈRE, 1792
Solid, obconic. Spire of medium height, coronate. Body whorl glossy-smooth, with nodulose shoulders, nearly straight sides, 7 to 10 widely spaced beaded cords present on the anterior half becoming obsolete posteriorly. Exterior of body whorl green-brown with a blue-white central spiral band, shoulder nodules white, spire white, interior purple, dark purple toward the anterior end.
  6 cm. Common. On rock and coral reefs. Indo-West Pacific; Rottnest Is., W.A. to northern N.S.W.

13, 13a. *Conus sanguinolentus* QUOY & GAIMARD, 1834
Solid, obconic. Spire of medium height, coronate. Body whorl with prominent shoulder nodules, slightly convex sides, anterior half with strongly beaded spiral cords, posterior half smooth. Spire pale green, nodules white, body whorl olive green, anterior beads on spiral cords white, anterior tip purple, interior purple.
  4.5 cm. Common. Shallow reefs. Indo-West Pacific; eastern Qld. This species is often misidentified as *C. lividus,* but the shell lacks the central white band of that species, the shoulder nodules and the anterior spiral cords are stronger, and the spire is pale green, whereas in *C. lividus* the spire is white.

14. *Conus parvulus* LINK, 1807
Moderately solid. Spire of medium height, coronate, with convex sides. Body whorl sculptured with fine spiral cords, shoulders angulate and nodulose. Anterior end of body whorl dark purple-brown, posterior end light grey-fawn with blue-grey spiral ribs spotted with brown, anterior and posterior colour zone separated by a narrow white or pale rose spiral band, spire blue-brown, nodules white, dark blue or brown blotches present between the nodules, interior blue-white.
  4.5 cm. Uncommon. N.T. and north Qld. This species is generally known as *C. imperator* WOOLACOTT, 1956 but Cernohorsky (1965) considers the Australian shells to be conspecific with Link's species from New Guinea.

Plate 100. Conidae (1⁵⁄₁₀× natural size)

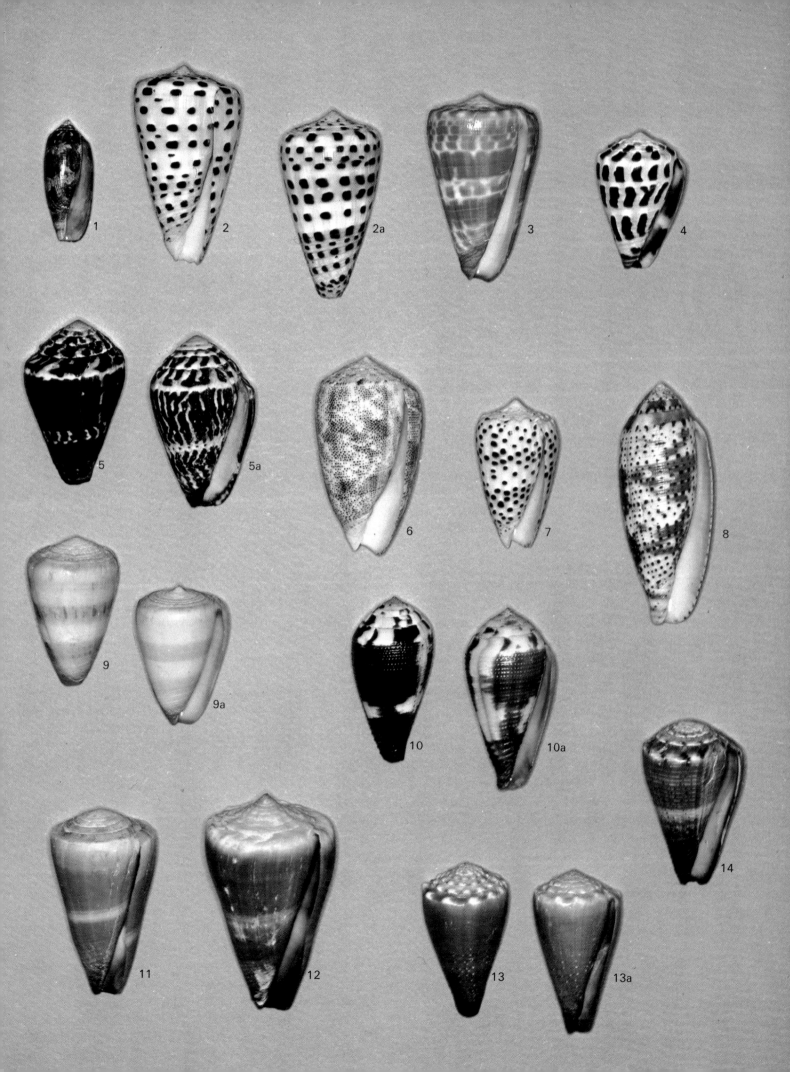

Conidae (Cont.)

1, 1a. *Conus sculletti* MARSH, 1962
Elongately obconic, with curiously concave sides. Spire low, shoulders sharply angulate. Body whorl smooth except for curved growth striae and very weak spiral cords at the anterior end. Exterior white, with irregular patches of tan or brown tending to concentrate in spiral bands, interior white.
 6 cm. Uncommon. Trawled off southern Qld.

2. *Conus rufimaculosus* MACPHERSON, 1959
Spire low, with a deep and wide spiral channel at the sutures. Shoulders angulate, sides of body whorl convex, surface minutely striate, weak spiral cords present at anterior end, body whorl otherwise smooth and glossy. Exterior pink-white, with irregular, confluent, triangular red-brown patches tending to form 3 spiral bands, deep interior pink, becoming white near the lip.
 5 cm. Moderately uncommon. Trawled off southern Qld. and northern N.S.W. The spiral channel at the sutures is an unusual and characteristic feature.

3. *Conus angasi* TRYON, 1884
Rather thin, obconic, broad and subangulate at the shoulders, sides of body whorl slightly convex. Spire of low to medium height, with a tiny papilliform apex. Spire and anterior end of the body whorl finely spirally striate, central and posterior part of the body whorl smooth. Aperture rather wide, columella almost straight with a terminal ridge at the anterior end bordering the anterior canal. Body whorl exterior white or cream, sometimes suffused with pink and overlain by irregular brown blotches, indistinct brown spiral bands and spiral rows of large transversely-elongated rectangular brown spots or bars; spire with radiating streaks of brown, interior white or pink.
 4 cm. Uncommon. Trawled in deep water off southern Qld. and N.S.W. This species resembles *C. advertex* but the spire is higher, the shoulders less angulate and the spiral bands less prominent.

4. *Conus advertex* GARRARD, 1961
Obconic, broad and angulate at the shoulders, sides of body whorl rather straight, spire flat with a small sharp projecting apex. Weak spiral striae present on the spire, body whorl smooth or with fine punctate striae which become spiral cords anteriorly. Exterior white or pink, with spiral rows of squarish brown spots of varying size, and 2 bands of light brown blotches, radiating streaks of orange-brown present on the spire, interior pink.
 4 cm. Uncommon. Trawled off southern Qld. and northern N.S.W. A striking, recently discovered species.

5. *Conus nielsenae* MARSH, 1962
Elongate, obconical, spire short or of medium height with prominently concave sides. Shoulder sharply angulate, sides of body whorl slightly concave, smooth except for a few shallow but rather wide anterior spiral grooves. Exterior polished cream or pale yellow, with thin slightly darker spiral lines, radial orange flecks present on the spire, interior pinkish.
 5.5 cm. Uncommon. Trawled at depths of about 20 fathoms off Townsville, eastern Qld.

6. *Conus vitulinus* HWASS [IN] BRUGUIÈRE, 1792
Obconic, spire low with a tiny apex and concave sides. Body whorl with angular shoulders and slightly convex sides, sculptured with beaded spiral cords anteriorly. Sutures sometimes grooved, strong spiral striae present on the spire. Exterior white, with 2 broad brown spiral bands and wavy axial streaks, anterior tip purple-brown; interior white.
 6 cm. Common on reefs. Indo-West Pacific; eastern Qld.

7, 7a. *Conus mitratus* HWASS [IN] BRUGUIÈRE, 1792
Solid, very slender, fusiform. Spire high with convex sides, shoulders very low and rounded, sides of body whorl slightly convex. Spire and body whorl sculptured with finely nodulose spiral cords which become coarser and further apart anteriorly. Exterior polished, cream or light rusty yellow with brown axial streaks and 3 broad brown spiral bands on the body whorl, interior bluish-white.
 3.5 cm. Uncommon. Indo-West Pacific; eastern Qld.

8. *Conus frigidus* REEVE, 1848
Solid, obconic. Spire spirally striate, of medium height with slightly convex sides. Body whorl with rounded shoulders and convex sides, sculptured with well spaced beaded spiral cords on the anterior half which become obsolete posteriorly. Exterior fawn to yellow-brown, sometimes with a slightly lighter spiral band at the shoulder and another at the centre of the body whorl, interior violet.
 5 cm. Uncommon. Under stones and coral. Indo-West Pacific; Barrow Is., W.A. to eastern Qld. The presence of anterior spiral cords and the coloration distinguish this species from *C. flavidus.*

9, 9a, 9b. *Conus monachus* LINNAEUS, 1758
Spire of medium height with straight or slightly convex sides. Shoulders of body whorl finely striate, rounded or slightly angulate. Body whorl with straight or slightly convex sides and numerous fine spiral cords which are strongest at the anterior end. Colour variable, usually blue-grey splashed with brown or green and white, with fine interrupted red-brown and white lines on the spiral cords, radial splashes of brown or green present on the spire, interior white.
 6 cm. Common. Under stones and in reef crevices. Central Indo-West Pacific; Jurien Bay, W.A. to eastern Qld. The figured specimens are from the Onslow area, W.A.; Pacific specimens are more heavily ribbed. A fish-eating species and likely to be dangerous. *C. achatinus* GMELIN, 1791 is a synonym.

10. *Conus tenellus* DILLWYN, 1817
Elongate, cylindrical, with slightly convex sides. Spire of medium height with concave sides, shoulders subangulate. Body whorl sculptured with rough crowded spiral cords. Exterior whitish with irregular light brown spiral bands, and small darker brown dashes on the spiral cords, interior white.
 5 cm. Uncommon. Central Indo-West Pacific and eastern Qld.

11. *Conus catus* HWASS [IN] BRUGUIÈRE, 1792
Solid, stout, spire of medium height with rather straight sides, body whorl with rounded shoulders and convex sides. Strong spiral striae present on the spire, spiral cords on the posterior part of the body whorl, and beaded spiral ribs anteriorly. Aperture rather wide. Exterior dark brown, with irregular white and sometimes bluish patches, anterior ribs spotted brown and white, interior blue-white.
 5 cm. Moderately common. On reefs. Indo-West Pacific; eastern Qld. A variable, fish-eating species and probably dangerous.

12, 12a. *Conus coronatus* GMELIN, 1791
Solid and stout. Spire coronate, strongly spirally striate, of medium height, sharply pointed, with slightly concave sides. Body whorl with angulate, nodulose shoulders and slightly convex sides, sculptured with fine beaded spiral cords which are strongest anteriorly. External colour variable, usually light bluish, grey, fawn or pink, with 2 darker broad spiral bands within which there may be dark brown blotches, 1 band near the centre and the other at the anterior end fine interrupted spiral lines of white and brown encircle body whorl, spire marked by brown to dark brown blotches between the shoulder nodules, deep interior blue-white becoming purple-brown near the margin.
 4 cm. Common. In sand pockets and among coral in shallow water. Indo-West Pacific; Pt. Cloates, W.A. to northern N.S.W.

13. *Conus rattus* HWASS [IN] BRUGUIÈRE, 1792
Moderately solid, obconic. Spire spirally striate, of medium height with almost straight sides. Body whorl with weakly angulate shoulders and straight sides, sculptured with crowded spiral cords which are coarse anteriorly but obsolete toward the shoulders. Exterior brown, with fine white spots and 2 spiral zones of large irregular blue-white blotches, 1 at the shoulder and 1 near the centre, spire white with radiating brown streaks, interior violet.
 5 cm. Common. In sand or in crevices and among corals Indo-West Pacific; Pt. Cloates, W.A. to eastern Qld.

14, 14a. *Conus magus* LINNAEUS, 1758
Moderately solid, elongate, sometimes subcylindrical. Spire strongly spirally striate, of medium height, sharply pointed, with slightly concave sides. Body whorl with subangulate to rounded shoulders and nearly straight sides, sculptured with numerous very fine spiral striae and beaded spiral cords near the anterior end. External colour very variable, usually white or cream, with fine interrupted spiral lines of dark brown dots and large patches of green, brown, olive-green or orange-yellow, irregular radiating brown lines present on the spire, interior white.
 6 cm. Common. Under stones and corals and in sand in shallow water. Indo-West Pacific; eastern Qld. Few cones can match this species for its variability of colour and pattern.

Plate 101. Conidae (1¹⁄₁₀× natural size)

auger shells

(FAMILY TEREBRIDAE)

Plate 102. *Terebra crenulata* LINNAEUS, 1758. At a depth of 6 feet, off Mossman, N. Qld. (*Photo*—Neville Coleman).

Q UEENSLAND COLLECTORS will be very familiar and perhaps a little dismayed with this family. Of the several hundred living species by far the greatest proportion live in the tropical waters of the Indo-West Pacific region. There are no cold-water augers although several species inhabit the temperate waters of southern Australia and New Zealand. Although auger shells are very attractive the family has not been reviewed in recent years and the number of species present on the shores of northern Australia is a matter for conjecture. Certainly there is a very large number in Queensland waters, but curiously the family is much less well represented on the tropical shores of northern Western Australia.

Many generic names have been used for members of the family Terebridae but their introduction has been piecemeal and no comprehensive classification has ever been proposed. We have used only the 3 broad generic units *Terebra*, *Duplicaria* and *Hastula*. Although some of the larger augers are well known there is a host of small species which are very difficult to identify. A selection of the more common Australian augers is illustrated in this book.

Augers (also known as "pencil" shells) take their vernacular name from their very long slender many-whorled shells which taper to a sharp tip. In general shape they resemble the shells of the family Turritellidae, but there is very little real relationship between the 2 families. In the Terebridae there is usually strong axial sculpture as well as spiral sculpture, there is a short anterior canal or notch, and the columella bears folds (plaits). In many species the columella is sharply twisted and there is a prominent anterior fasciole behind the anterior canal. The long slender shells of terebrids are adapted for the active predatory habits of the burrowing animals.

Some augers have prominent eye stalks but others are without eyes. The foot is small considering the size of the shell. There is a small operculum. There is very little knowledge on the biology of augers. Information on breeding, eggs and larval development is lacking, although the anatomy of the reproductive system indicates that fertilization is internal and the females lay their eggs in capsules.

It is known that, like their close relatives the cones, many augers have a poison gland and harpoon-like radular teeth with which they kill their prey. Marine worms are the most likely prey but there is little information about this. In a few species a radula and a poison gland are lacking and it is believed that these augers catch their prey by suction (Rudman, 1969). There is no evidence that the poison of those species which have a poison gland and radula is in any way dangerous to humans.

All augers live in sand. In Queensland and other coral reef areas a variety of species may be found by "tracking" or "fanning" the sand among coral heads in shallow pools and lagoons. Some species may be found on sand cays at low tide. Several species of the genus *Hastula* live on the surf-swept edges of beaches.

Selected references:

BURCH, R. D. (1964). Notes on the Terebridae of the Philippine Islands. Veliger, **6** : 210-218. (1965). New terebrid species from the Indo-Pacific Ocean and from the Gulf of Mexico, with new locality records and provisional list of species collected in Western Australia and at Sabah, Malaysia. Veliger, **7** : 241-253, pl. 31.

CERNOHORSKY, W. O. & JENNINGS, A. (1966). The Terebridae of Fiji. Veliger, **9** : 37-67, pls 4-7.

CERNOHORSKY, W. O. (1969). List of type specimens of Terebridae in the British Museum (Natural History). Veliger, **11** : 210-222.

RUDMAN, W. B. (1969). Observations on *Pervicacia tristis* (DESHAYES, 1859) and a comparison with other toxoglossan gastropods. Veliger, **12** : 53-63.

1, 1a. *Terebra guttata* RÖDING, 1798

Large, solid, whorls smooth but thickened at the shoulders. Base of the body whorl heeled. Aperture rather rectangular, parietal wall almost vertical to the columella. Pink, orange or orange-brown, with a pre-sutural band of large elongate white spots, and another around the anterior end of the body whorl.

14 cm. Moderately uncommon. Indo-West Pacific; north-eastern Qld.

2, 2a. *Terebra areolata* LINK, 1807

Large, rather thin, broad at the anterior end. Aperture elongate, tear-drop shaped. Early whorls faintly plicate, later whorls smooth except for a weak spiral pre-sutural groove. Cream or fawn, with spiral bands of large squarish brown spots, 4 bands visible on the body whorl, 3 on the other whorls.

6 cm. Common. Indo-West Pacific; Pt. Cloates, W.A. to eastern Qld. This species is often known as *T. muscaria* LAMARCK, 1822 but Link's name has priority.

3, 3a. *Terebra dimidiata* LINNAEUS, 1758

Large, aperture elongate, whorls smooth with an incised spiral pre-sutural line. Orange or orange-red, with white irregular axial lines in the zone anterior to the pre-sutural line, crossed by 1 or 2 weaker white spiral lines.

13 cm. Common. Indo-West Pacific; Pt. Cloates, W.A. to central Qld.

4, 4a. *Terebra subulata* LINNAEUS, 1767

Large, narrow, aperture tends to be rectangular, parietal wall almost vertical to the columella. Base of the body whorl heeled.

Early whorls with a faint spiral pre-sutural groove and weak axial ribs, later whorls smooth except for strong growth striae and slightly thickened shoulders anterior to the sutures. A broad spiral channel present behind the anterior canal. Cream, with spiral bands of large dark brown spots, 3 bands visible on the body whorl, 2 on the other whorls.

18 cm. Common. Indo-West Pacific; Pt. Cloates, W.A. to eastern Qld.

5, 5a. *Terebra crenulata* LINNAEUS, 1758

Large, rather solid, broad at the anterior end. Base of the body whorl heeled. Whorls smooth usually with a spiral row of pointed white crenules or knobs anterior to the sutures. Flesh or fawn with darker zones, irregular short brown axial lines arising from the sutures, spiral rows of brown dots, 3 rows visible on the body whorl, 2 on the other whorls.

15 cm. Common. Indo-West Pacific; Pt. Cloates, W.A. to eastern Qld.

6, 6a. *Terebra maculata* LINNAEUS, 1758

Large, solid and heavy, with an elongate aperture. Base of the body whorl weakly heeled. Early whorls weakly plicate, later whorls smooth. Cream, with spiral band of squarish light tan patches, and 2 spiral bands of squarish dark brown patches close to the posterior margin of each whorl.

24 cm. Common. Indo-West Pacific; Dampier Archipelago, W.A. to eastern Qld. Although common elsewhere this species is rare in W.A.

Plate 103. Terebridae ($\%_{10}$x natural size)

Terebridae (Cont.)

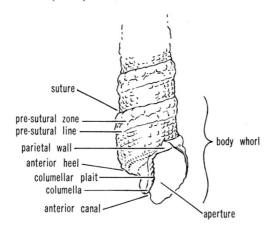

suture

pre-sutural zone
pre-sutural line

parietal wall

anterior heel

columellar plait

columella

anterior canal

body whorl

aperture

Fig. 34. A typical Auger showing major external parts of shell.

1. *Terebra chlorata* LAMARCK, 1822
Solid, rather short and broad. Early whorls with very weak plications, later whorls smooth with a weak spiral pre-sutural groove. White or cream, with 4 spiral bands of diffuse purple-brown blotches or wavy lines on the body whorl, bands indistinct on higher whorls.
   9 cm. Common. Indo-West Pacific; northern Qld.

2. *Terebra hectica* LINNAEUS, 1758
Solid, rather short and broad, left side of the body whorl almost straight. Narrow low spiral callus present immediately behind the sutures. Whorls smooth and polished. Colour variable, usually creamy-white, with an interrupted or continuous irregular brown and yellow band at the sutures which becomes broader toward the apex, and may there cover the whorls completely.
   9 cm. Moderately uncommon. Indo-West Pacific; N.T. and northeastern Qld. Often known by the name *T. caerulescens* LAMARCK, 1822.

3. *Terebra felina* DILLWYN, 1817
Solid, rather short and broad. Whorls smooth, except for a faint spiral pre-sutural groove and faint axial ribs on the early whorls. White, with 2 spiral rows of small irregular brown spots on the body whorl, only 1 row visible on other whorls.
   9 cm. Uncommon. Indo-West Pacific; Pt. Cloates, W.A. to eastern Qld.

4. *Terebra lima* DESHAYES, 1859
Long and moderately narrow with many whorls, aperture almost rectangular. Base of the body whorl strongly heeled. Whorls strongly sculptured by intersecting spiral and axial ribs forming a cancellate pattern, the first spiral rib anterior to the pre-sutural zone stronger than the others. Pre-sutural groove wide and deep. White to yellow.
   10 cm. Moderately common. Indo-West Pacific; eastern Qld. and N.S.W. The figured specimen was trawled in 25 fathoms off Trial Bay, N.S.W.

5. *Terebra commaculata* GMELIN, 1791
Long and narrow with many whorls, sutures grooved. Base of the body whorl strongly heeled. Whorls with 2 heavy nodulose spiral ribs at the sutures, separated by a strong incised spiral line, and followed by several spiral rows of fine nodules. Aperture almost rectangular, parietal wall almost vertical to the columella. White with elongate squarish brown patches, and a spiral row of round brown spots at the anterior end of the body whorl.
   8 cm. Uncommon. Northern Indian Ocean and central Indo-West Pacific; recently trawled at 50 fathoms off the Kimberley region, W.A.

6. *Terebra albomarginata* DESHAYES, 1859
Solid, elongate with many whorls. Aperture almost rectangular, base of body whorl heeled. Pre-sutural groove deeply incised. Pre-sutural zone with a thick rounded raised band which is crossed by numerous low ribs, ribs continue as weak curved axial lirae on the anterior parts of the whorls, axial lirae crossed by strong spiral ribs forming a cancellate pattern. Whorls orange, pre-sutural zone white.
   7 cm. Moderately uncommon. Indo-West Pacific; eastern Qld. as far south as Southport.

7. *Terebra nebulosa* SOWERBY, 1825
Solid, aperture elongate, base of body whorl heeled. Early whorls with broad axial ribs which become obsolete on later whorls. Whorls also sculptured by a deep punctate pre-sutural spiral groove and numerous fine spiral striae. Exterior white with diffuse orange-red blotches and thin spiral lines, aperture orange.
   9 cm. Uncommon. Indo-West Pacific; Shark Bay, W.A.

8. *Terebra triseriata* GRAY, 1834
Long and very slender, with very many whorls. Aperture tends to be rectangular. Base of the body whorl strongly heeled. Whorls sculptured by 2 thick juxtaposed nodular spiral ribs at the sutures, and several spiral rows of small nodules. Fawn or red-brown.
   11 cm. Uncommon. Central Indo-West Pacific; Onslow, W.A. to eastern Qld.

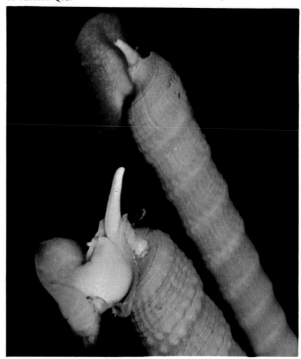

Plate 104. *Terebra triseriata* GRAY, 1834, with animal partially extended, Qld.   (*Photo*—Don Byrne).

9, 9a, 9b, 9c. *Duplicaria duplicata* LINNAEUS, 1758
Moderately slender with many whorls. Sides of whorls almost straight. Base of body whorl weakly heeled. Early whorls with weak axial plications which become broad flat axial ribs separated by incised axial lines on later whorls. Pre-sutural spiral groove deeply incised, pre-sutural zone wide and flat. Colour variable, may be blue-grey with elongate brown patches and a light brown spiral band ((9c) a specimen from Pt. Cloates, W.A.), commonly a uniformly chocolate brown ((9b) a specimen from Dampier Archipelago, W.A.), or sometimes uniformly pale pink or orange ((9, 9a) specimens from Maud Landing, W.A.). A light thin spiral line usually present on the anterior part of the body whorl.
   9 cm. Common. Central Indo-West Pacific; Shark Bay, W.A. to N.T.

10, 10a, 10b. *Duplicaria evoluta* DESHAYES, 1859
Moderately slender with many whorls. Base of the body whorl weakly heeled. Sides of whorls slightly convex. Pre-sutural groove punctate, deep and wide, pre-sutural zone narrow and rounded. Whorls sculptured by many narrow angular axial ribs interrupted by the pre-sutural groove. Colour variable, usually uniformly cream or brown with a pale spiral line on body whorl, sometimes fawn, blue-grey or black.
   9 cm. Common. Central Indo-West Pacific; Shark Bay to Broome, W.A. *D. australis* SMITH, 1873 is like this, but the pre-sutural groove is said to be smooth, not punctate. The two may be forms of the one species. *D. evoluta* occurs in the same areas in W.A. as *D. duplicata* and is often mistaken for that species, but may be distinguished by the narrow angular ribs and the narrow rounded pre-sutural zone.

Plate 105. Terebridae ($1\frac{2}{10}$x natural size)

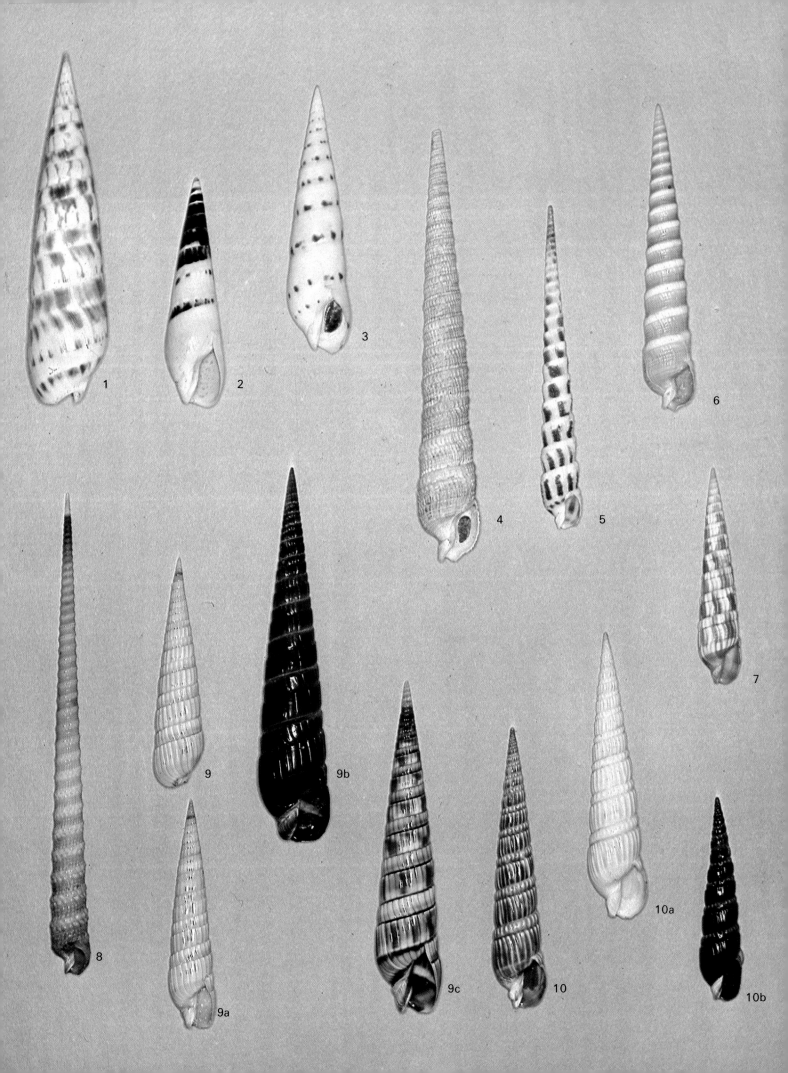

Terebridae (Cont.)

1. *Terebra babylonia* LAMARCK, 1822
Elongate, with many whorls. Base moderately heeled, aperture rather rectangular. Whorls rough, sculptured by an incised spiral pre-sutural groove, 2 weaker anterior spiral grooves, many spiral striae and many weak curved axial grooves dividing the surface into flat squarish nodules. Fawn or flesh-coloured with orange-brown grooves and a broad spiral orange-brown band anteriorly, interior orange-brown.
10 cm. Common. Indo-West Pacific; Pt. Cloates, W.A. to eastern Qld.

2. *Terebra pertusa* BORN, 1778
Elongate with many whorls. Base of body whorl heeled; whorls with rather straight sides. Aperture almost rectangular. Sutures incised, pre-sutural groove deep. Pre-sutural zone narrow, crossed by straight white axial ribs separating small rectangular patches of white or dark brown, remainder of whorls with spiral grooves interrupted by curved axial ribs, and many spiral striae around the base at the anterior end. Yellow or pale orange with a white spiral band at anterior shoulder, white ribs, and the aforementioned brown markings at sutures.
8 cm. Uncommon. Indo-West Pacific; eastern Qld.

3. *Duplicaria bernardi* DESHAYES, 1857
Moderately elongate with many whorls, sutures deeply grooved. Whorls rather long, sculptured by numerous straight, narrow, axial ribs interrupted by the pre-sutural groove. Greenish-grey with a central pale yellow spiral band, a red-brown spiral band around the anterior canal and a red-brown apex.
4 cm. Common. Western Pacific; eastern Qld. to northern N.S.W.

4. *Duplicaria crakei* BURCH, 1965
Small, sutures deeply incised. Pre-sutural groove wide, deep. Whorls with strong, narrow rather straight axial ribs interrupted by the pre-sutural groove, interstices between axial grooves smooth, not striate. Posterior ends of the ribs form prominent elongate crenules in the pre-sutural zone. Early whorls amber-coloured, later whorls usually pale yellow sometimes blotched with brown, pre-sutural groove and a broad spiral band at the anterior margin are blue-purple, pre-sutural zone sometimes blue-grey.
2.5 cm. Common. Broome, W.A. This species resembles *D. addita*, differing by its less rounded and less shouldered pre-sutural band, and its deeper and wider pre-sutural groove.

5. *Terebra affinis* GRAY, 1834
Moderately elongate, base of the body whorl weakly heeled. Pre-sutural groove deep and punctate. Pre-sutural zone narrow and forms a thick spiral rib. Whorls with irregular discontinuous axial grooves and fine spiral striae, on early whorls the axial grooves separating strong axial ribs. White with orange grooves, elongate orange-brown patches, and central and anterior spiral bands of pale orange bounded by spiral orange lines.
6 cm. Common. Indo-West Pacific; Pt. Cloates, W.A. to eastern Qld.

6. *Terebra cerethina* LAMARCK, 1822
Solid, rather stout. An elevated parietal columellar callus present. Early whorls sculptured by many strong rounded axial ribs which are absent or weak on the later whorls. Pre-sutural groove weak, especially on the body whorl. White to blue-grey background, overlain with a light orange-yellow stain, and with 4 darker orange-yellow spiral lines and irregular axial lines.
8 cm. Common. Indo-West Pacific; eastern Qld.

7. *Hastula lauta* PEASE, 1869
Resembles *H. strigilata*, but longitudinal ribs stronger, fewer, and further apart, and obsolete on the anterior ⅓ of the body whorl. Usually green-grey rather than blue-grey.
3 cm. Common. Central Indo-West Pacific including Hawaii, Broome, W.A. to north-eastern Qld. The illustrated specimen is from Hawaii.

8, 8a. *Hastula strigilata* LINNAEUS, 1758
Small, slender, aperture elongate, anterior end drawn out. A very indistinct pre-sutural groove sometimes present. Whorls sculptured by strong axial ribs which continue to the anterior end of the body whorl. Blue-grey with a white or pale grey spiral band containing a single row of large brown spots in front of the sutures, and usually a narrow white spiral line nearer the anterior end, columella white.
4 cm. Common. Western Pacific; eastern Qld. There are several species like this, including *H. lauta* and *H. rufopunctata*, in the Indo-West Pacific region. They inhabit the edge of sandy beaches swept by light to moderate surf. Because of their variability and similarity they are sometimes difficult to tell apart. In fact, future biological studies may reveal that some of the "species" now recognized in this complex are merely variants.

9. *Terebra albida* GRAY, 1834
Solid, body whorl broad and rounded, base heeled. Aperture wide, more or less oval. Whorls smooth, except for irregular oblique growth striae. Pre-sutural groove represented by a shallow spiral depression in front of the sutures. Most often uniformly white or cream, but sometimes with a row of small brown spots in front of the sutures.
4 cm. Uncommon. Vic., Tas. and S.A. The genus *Nototerebra* was erected for this southern species by Cotton (1947).

10. *Duplicaria addita* DESHAYES, 1859
Small, sutures incised. Pre-sutural groove incised, narrow. Whorls with strong narrow rather straight axial ribs interrupted by the pre-sutural groove, interstices between the axial ribs smooth, not striate. Early whorls dark purple, later whorls blue-grey with occasional large blue-brown spots and a pale central spiral band, pre-sutural zone also often pale.
2.5 cm. Common. Fremantle to Broome, W.A.

11, 11a, 11b. *Hastula rufopunctata* E. A. SMITH, 1877
Resembles *H. strigilata*, but with axial ribs absent, or obsolete and confined to the posterior ½ of the body whorl. Colour most often like *H. strigilata*, but commonly fawn or cream.
3 cm. Abundant in restricted localities. Shark Bay to Broome, W.A.

12. *Terebra undulata* GRAY, 1834
Solid, moderately elongate with many whorls, base of body whorl heeled. Pre-sutural groove indistinct, pre-sutural zone contains a spiral row of large raised squarish nodules. The whorls anterior to the pre-sutural groove sculptured by broad curved axial ribs with numerous fine transverse striae in the grooves between them. Pre-sutural nodules white, axial ribs orange, grooves orange-brown.
6 cm. Moderately uncommon. Western Pacific, including eastern Qld.

13. *Hastula nitida* HINDS, 1844
Small, body whorl drawn out and narrowed anteriorly, outer lip flared outwards. Aperture elongate, columella inclined but almost straight. Pre-sutural groove absent, whorls with well spaced distinct, strong, straight axial ribs confined to the posterior portion of the body whorl and obsolete toward the flared lip. Blue-grey with a white spiral line at the sutures or sometimes uniformly fawn.
3 cm. Common. Indo-West Pacific; Rottnest Is., to Broome, W.A.

14. *Hastula lanceata* LINNAEUS, 1767
Solid, slender, glossy smooth. Early whorls with straight axial ribs which become weak or obsolete on the last 3-4 whorls (where they are easier to feel than to see). Pre-sutural groove absent. Shiny white or cream with thin widely spaced and rather straight red-brown axial lines.
7 cm. Uncommon. Indo-West Pacific; eastern Qld. For this species Dall (1908) introduced the generic name *Acuminia*.

15. *Hastula brazieri* ANGAS, 1871
Small, moderately elongate. Pre-sutural groove absent or indistinct. Whorls glossy smooth and without sculpture, except for a spiral row of weak axially-elongated nodules in front of the sutures. Cream with irregular longitudinal streaks of red-brown. On the body whorl a clear cream spiral band near the anterior end and a brown spiral band between this and the anterior extremity.
3 cm. Uncommon. N.S.W. to S.A. Some authors place this species in the genus *Acuminia* along with *lanceata*.

Plate 106. Terebridae (1⁸⁄₁₀ˣ natural size) ⬎

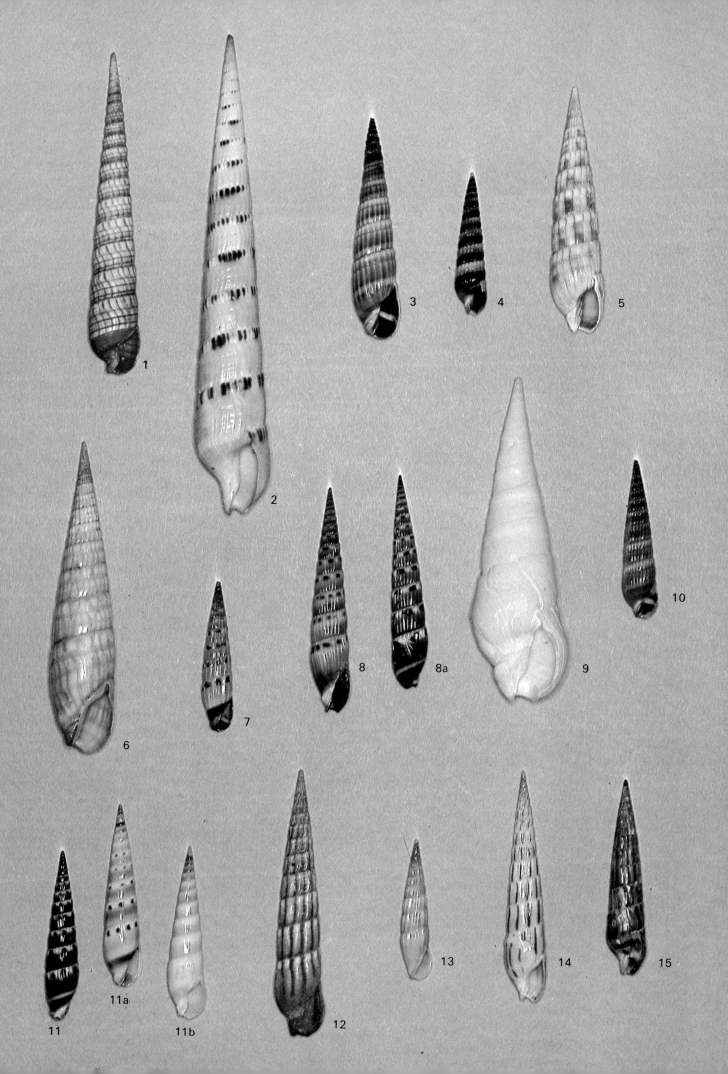

ACUMINATE	Tapered to a point
ANTERIOR CANAL	Notch or trough-like or tubular extension of anterior apertural margin supporting the incurrent siphon (=siphonal canal)
APERTURE	Opening or entrance of the shell providing outlet for the head and foot
APEX	First-formed tip of the shell (plural apices)
APICAL WHORLS	Those whorls near the apex
AXIAL	Parallel, or nearly so, with the shell axis (=longitudinal)
AXIS	Imaginary line through the apex, about which the whorls are coiled
BENTHIC	Bottom dwelling
BICONIC	Resembling two cones placed base to base
BODY WHORL	Last and usually the largest whorl of the coiled shell
CALLUS	Calcareous thickening (=callosity)
CANCELLATE	Ornamented with intersecting spiral and axial ridges
COLUMELLA	Pillar surrounding the axis of a coiled shell, formed by the inner walls of the whorls
COLUMELLAR LIP	Inner edge of the aperture comprising the visible part of the columella
CONSPECIFIC	Of the same species
CONVOLUTE	With last whorl completely concealing the earlier ones
CORD	Fine round-topped spiral or axial ridge
CORONATE	With tubercles or nodules at the shoulders of the whorls
COSTA	Rib (plural costae)
CRENATE	With the edge notched
DEXTRAL	Shell coiled in a right-hand spiral, i.e. clockwise when viewed from the apex
ENDEMIC	Peculiar to a region
FASCIOLE	Spiral band formed by successive edges of a canal or notch (either anterior or posterior)
FOSSULA	Shallow linear depression of the anterior part of the inner lip of some cowries
FUNICLE	Ridge spiralling into the umbilicus, as in Naticidae
FUSIFORM	Spindle-shaped, swollen at the centre and tapering almost equally toward the ends
GENUS	Group of genetically related species possessing certain characters in common and easily separable from other groups of species (plural genera)
GROWTH LINES	Lines which represent former resting positions of the outer lip
HEIGHT	Distance along the axis between the apex and the anterior end of the shell (except in cowries)
HOLOTYPE	The single specimen designated as "the type" by the original author of a species name at the time of the original description
HOMONYM	One of two or more identical but independently proposed names for different species, genera or other taxa
IMBRICATED	With laminae overlapping each other like tiles
IMPRESSED	Indented
INNER LIP	Edge of the aperture near the axis, extending from the suture to the anterior end of the columella (=labium)
INTERSPACES	Grooves between ribs or cords (=interstices)
LAMELLA	Thin plate (plural lamellae) = lamina (plural laminae)
LIRA	Linear elevation (ridge) on the shell surface within the outer lip (plural lirae)
LITTORAL	The tidal zone of the shore, i.e. between high and low tide levels
MACULATE	Patterned with blotches, i.e. with maculations
MAMMILLATED	With the protoconch rounded like a teat, or with dome-shaped protuberances on the shell surface
MULTISPIRAL	With many whorls
NODULOSE	With numerous or conspicuous nodules (=nodose)
NOMINATE	Subspecies (or subgenus) having the same name as the subdivided species (or genus) and the same type

NUCLEUS	Earliest formed part of a shell or operculum
OBCONICAL	With the form of an inverted cone
OPERCULUM	Horny or calcareous structure borne by the foot and usually serving for the closure of the aperture
OUTER LIP	Outer margin of the aperture=labrum
PARATYPE	A specimen other than the holotype which was before the author at the time of preparation of the original description of a species and was so designated or indicated by the author
PARIETAL REGION	That part of the body wall just within and just without the aperture, and posterior to the columella
PELAGIC	Inhabiting the open ocean
PENULTIMATE	Next to the last formed; refers to the last whorl but one
PERIOSTRACUM	Outer layer of horny material covering the calcareous shell
PILLAR STRUCTURE	Calloused part of the body whorl to the left of the columella
PLAIT	Spiral fold or ridge on the columellar lip (=plication)
PLANISPIRAL	Coiled in a single plane
PLANKTONIC	Drifting or weakly swimming in the ocean
PLICATE	Folded or twisted or bearing plaits
POSTERIOR CANAL	Notch, or trough-like or tubular extension of the posterior apertural margin supporting the excurrent siphon
PROTOCONCH	Embryonic shell, present in the adult as the apical or nuclear whorls and often clearly demarcated from later whorls by change of sculpture
PUNCTATE	Minutely pitted
PYRIFORM	Pear-shaped
RECURVED	With the distal end (e.g. anterior canal) bent away from the observer when the shell is viewed from the apertural side
REFLECTED	Turned outward and backward at the margin
RETICULATE	Forming a network of intersecting lines
RIB	Round-topped elevated ridge of moderate width and prominence
RUGOSE	Rough
SERRATE	Notched or toothed at the edge like a saw
SINISTRAL	Shell coiled in a left-hand spiral, i.e. anticlockwise when viewed from the apex
SPECIES	Groups of actually or potentially interbreeding natural populations which are reproductively isolated from other such groups
SPIRE	That part of the shell consisting of all the whorls except the last
SQUAMOSE	With scales
STRIA	Narrow and shallow incised groove (plural striae)
SUBLITTORAL	Below low tide level
SUBSPECIES	A geographically defined group of populations comprising individuals which differ in certain characters from other such subdivisions of the species
SULCUS	Groove or furrow (plural sulci)
SUTURE	Continuous line on shell surface where whorls adjoin
SYNONYM	Each of two or more different names for the same species, genus or other taxon
TAXON	A taxonomic group that is sufficiently distinct to be distinguished by name (plural taxa)
TAXONOMY	The theory and practice of classifying organisms
TYPE	see HOLOTYPE
TYPE-LOCALITY	Locality at which the type specimen was collected
TYPE SPECIES	The species which was designated type of a genus
UMBILICUS	Cavity or hollow around the axis of shells formed because the inner walls of the whorls do not meet to form a solid columella
VARIX	Elevated ridge formed by a thickened and reflected former lip (plural varices)
WHORL	Any complete coil of a spiral shell

Numbers in bold type refer to plates.
Numbers in brackets after plate numbers refer to shell numbers on the plates.
Page numbers and text-figure numbers are in Roman type.

drilling—by carnivores : 14, 64, 83, 90
Drupa Röding 1798 : 92
Drupella Thiele 1925 : 94
Drupes : 90
Drupina Dall 1923 : 92
duffusi Iredale 1936, **Pterynotus** : 88, 88 fig. 18, **59**(8)
Dulcerana Iredale 1931, subgen. of **Colubrellina** : 80
Duplicaria Dall 1908 : 154, 156, 158
duplicata Linnaeus 1758, **Duplicaria** : 156, **105**(9, 9a, 9b, 9c)
dwarf forms : 45, 70
dwarf tritons : 81 plate 54, 82

ear shells : 14, 15, 24
Eastern Overlap Zone : 17, 18 fig. 10
ebeninus Bruguière 1792, **Pyrazus** : 32, **12**(10, 10a)
ebraeus Linnaeus 1758, **Conus** : 150, **100**(4)
eburnea Reeve 1846, **Cominella** : 96, **63**(7, 7a)
eburnea Barnes 1824 = **Cypraea** *miliaris* : 52, **32**(8a)
eburneus Hwass 1792, **Conus** : 150, **100**(2, 2a)
echinata Blainville 1832, **Thais** : 90, **60**(7, 7a)
Echinodermata : 66
echinoderms—as food : 14, 72
—as host : 14
Echinoidea : 66
Ectosinum Iredale 1931 : 64
egg capsules : **4**, 45, 66, **47**, 76, 80, 82 fig. 17, 83, 96, 102, 105, 114, 120, 121, 140, **95**, 144 fig. 31, 148, 154
egg collars : 64
egg cowrie : 60-63
egg mass : 26, 28, 30, 32, 36, 45, 64, 66, 96, 120, 121, 122, 140
eggs : 14, 15, **6**, 24, **26**, 66, 120, 154
eglantina Duclos 1833, **Cypraea** : 50, **29**, **30**(1, 1a)
elegans Philippi 1899, **Haliotis** : 24, **8**(5, 5a)
elongata Gray 1847, "**Amalda**" : 105, 108, **71**(7, 7a)
Ellatrivia Iredale 1931 : 62
ellioti Sowerby 1864, **Amoria** : 121, 128, **85**(7)
emblema Iredale 1931, **Cypraea** : 43
embryo : 15 fig. 7, 83, 105, 120, 121, 140
emmae Reeve 1846, **Haliotis** : 24, **8**(3)
Eocithara Fischer 1883 : 110
epidromis Linnaeus 1758, **Strombus** : 38, **17**(2)
episcopalis Lamarck 1810 = **Oliva** *caerulea* : 106
episcopus Hwass 1792, **Conus** : 146, **98**(1)
episema Iredale 1939 = **Cypraea** *venusta* : 46
Epitoniidae : 34
Epitonium Röding 1798 : 34
eremitarum Röding 1798, **Mitra** : 114, **74**(4, 4a)
Ericusa H. & A. Adams 1858 : 122, 124
erinacea Linnaeus 1758, **Casmaria** : 70, **49**(7, 7a)
erosa Linnaeus 1758, **Cypraea** : 52, **32**(3)
errones Linnaeus 1758, **Cypraea** : 52, **32**(5)
erythrinus Dillwyn 1817, **Strombus** : 40, **18**(5)
erythrostoma Meuschen 1789 = **Oliva** *miniacea* : 106
estuaries : 102
Ethminolia Iredale 1924 : 16
euclia Hedley 1914, **Charonia** : 76
euclia Steadman & Cotton 1946, **Cypraea** : 43
Eumitra Tate 1889 : 118
Euprotomus Gill 1870, subgen. of **Strombus** : 38
evoluta Deshayes 1859, **Duplicaria** : 156, **105**(10, 10a, 10b)
exasperatum Gmelin 1791, **Vexillum** : 116, **75**(14, 14a)
eximius Perry 1811, **Phasianotrochus** : 26, **9**(10, 10a, 10b, 10c)
exoptanda Reeve 1849, **Amoria** : 124, **81**(5)
exquisita Iredale 1931, **Austroharpa** : 110, **72**(6, 6a)
extraneus Iredale 1936 = **Chicoreus** *denudatus* : 86

extrema Iredale 1939, subsp. of **Cypraea** *clandestina* : 56

facifer Iredale 1935, subsp. of **Cypraea** *limacina* : 58
fallax Smith 1881, subsp. of **Cypraea** *cribraria* : 52
fan corals : 60
fasciatum Bruguière 1792, **Cerithium** : 32, **12**(3, 3a, 3b, 3c, 3d, 3e)
Fasciolariidae : 98-101
faunal provinces : 17, 18
faunal regions : 17, 18 fig. 10
feeding types—carnivores : 43, 66, 72, 83, 90, 96, 98, 110, 112
—detrital feeders : 14
—filter feeders : 14
—herbivores : 14, 15, 26, 28, 30, 32, 36
—parasites—molluscs : 14
—flat worm : 32
—predators : 14, 20, 43, 64, 66, 72, 76, 90, 96, 105, 114, 120, 121, 140, 142, 154
—scavengers : 96, 105, 114
feeding mechanisms—teeth and radula : 14, 14 fig. 5, 45, 136 fig. 28, 141 fig. 30b
—drilling : 14, 64, 83, 90
—acid secretions : 64, 76, 83
—sucking : 94
—ingestion : 66
—digestion : 64
felina Gmelin 1791, **Cypraea** : 56, **35**(6, 6a)
felina Dillwyn 1817, **Terebra** : 156, **105**(3)
fertilization—external : 14, 24, 26
—internal : 15, 154
ferruginea Lamarck 1811, **Mitra** : 114, **74**(6, 6a)
Ficidae : 2 frontpiece, 16 fig. 8, **49**, 72
ficoides Lamarck 1822, **Ficus** : 72
ficus Linnaeus 1758, **Ficus** : 16 fig. 8, 72
Ficus Röding 1798 : 2 frontpiece, 16 fig. 8, **49**, 72
Fig shells : 2 frontpiece, **49**, 72
figulinus Linnaeus 1758, **Conus** : 140, 144, **97**(4)
filamentosa Röding 1798, **Pleuroploca** : 98, **64**(3, 3a)
filaris Linnaeus 1771, **Cancilla** : 118, **76**(13)
filosa Sowerby 1892, **Ficus** : 72
filter feeding : 14
fimbriata Quoy & Gaimard 1833, **Cassis** : 68, **48**(1, 1a, 1b)
fimbriata Gmelin 1791, **Cypraea** : 56, **35**(10)
fimbriatus Lamarck 1822, **Murexsul** : 88
fish—as prey : 14, 140, **94**, 142, 144, 148, 150, 152
—shells in stomachs of : 45
fisheries—abalone : 24
—overfishing : 20, 21
flammeum Röding 1798 = **Melo** *amphora* : 138
flat-worm as parasite : 32
flavicans Gmelin 1791, **Aulica** : 132, **88**(1, 1a)
flavidus Lamarck 1810, **Conus** : 150, **100**(11), 152
flindersi Verco 1914, **Altivasum** : 112, **73**(3)
flindersi Adams & Angas 1863, **Lepsiella** : 92, **61**(11)
Flindersian province : 18
food—algae : 14, 26, 28, 30, 32, 36
—anemones : 34, 94
—ascidians : 76
—barnacles : 90
—bivalves : 16, 64, 90, 102
—coelenterates : 14, 34, 60, 62, 90, 94
—echinoderms—echinoids : 14, 66, 72
—starfish : 20, 76
—fish : 14, 140, **94**, 142, 144, 148, 150, 152
—gorgonians : 60, 62
—hydrozoans : 94
—invertebrates in general : 14
—molluscs : 14, 76, 83, 90, 96, 98, 140, 142
—plankton : 14, 94
—phytoplankton : 15
—soft corals : 60, 94

—sponges : **5, 7**, 43, 45, **23, 24**
—worms : 14, 80, 140, 154
francolina Bruguière 1789, **Nassa** : 94, **62**(3)
free-swimming larvae : 66, 105, 120, 121
fressa Iredale 1933 = **Architectonica** *perspectiva* : 34
friendii Gray 1831, **Cypraea** : **5**, 14 fig. 5, 42, 45, 46, **25**(1, 1a, 1b, 1c, 1d, 1e, 1f, 1g, 2, 2a)
frigidus Reeve 1848, **Conus** : 152, **101**(8)
Frog shells : 80
fulgetrum Sowerby 1825, **Ericusa** : 124, **81**(2, 2a)
Fusinidae : 98
Fusinus Rafinesque 1815 : 98
Fusitriton Cossmann 1903 : 78

Galeodidae : 96
gedlingae Cate 1968, subsp. of **Cypraea** *leviathan* : 50
gemmatum Reeve 1844, **Cymatium** : 76
gemmulata Lamarck 1822, **Niotha** : 102, **66**(10, 10a)
generalis Linnaeus 1767, **Conus** : 146, **98**(7, 7a)
geographus Linnaeus 1758, **Conus** : **94**, 142, 144, **97**(1), 150
georginae Gray 1834, **Melo** : 138, **92**(2)
Giant whelks and conchs : 96, **63**, 98
Gibberulus Jousseaume 1888, subgen. of **Strombus** : 40
gibberulus Linnaeus 1758, **Strombus** : 40, **18**(9, 9a)
gibbosus Röding 1798, subsp. of **Strombus** *gibberulus* : 40, **18**(9, 9a)
gibbulus Gmelin 1791, **Latirus** : 100, **65**(7)
gigantea Smith 1914, form of **Tutufa** *bubo* : 80
gigas McCoy 1867, **Cypraea** : 42
glabra Swainson 1821, **Eumitra** : 118, **76**(2)
glabratum Dunker 1852, **Phalium** : 68, **48**(10, 10a)
glands : 83—mucous : 12
—venom : 140, 154
—salivary : 66, 76
glans Linnaeus 1758, **Alectrion** : 102, **66**(11, 11a)
glans Hwass 1792, **Conus** : 148, **99**(1)
glaucum Linnaeus 1758, **Phalium** : 66, 68, **48**(3, 3a)
globulus Linnaeus 1758, **Cypraea** : 62, **44**(1, 1a)
gloriamaris Chemnitz 1778, **Conus** : 96
gracilis Gaskoin 1849, **Cypraea** : 56, **35**(12, 12a)
gracilis Broderip & Sowerby 1829, **Harpa** : 110
gorgonians : 60, 62
Grandeliacus Iredale 1957 : 34
grandis Gray 1839, **Penion** : 96, **63**(4)
grangeri Sowerby 1900, ? = **Lyria** *mitraeformis* : 126, **83**(7b)
granifera Kiener 1834, **Plicarcularia** : 102, **66**(6, 6a)
granosa Quoy & Gaimard 1834, **Granuliscala** : 34, **13**(1, 1a)
granti Pritchard & Gatliff 1902 = **Haliotis** *conicopora* : 24
granularis Röding 1798, **Colubrellina** : 80, **54**(7, 7a, 7b)
granulata Duclos 1832, **Morula** : 92
Granuliscala Boury 1909 : 34
grayi Ludbrook 1953, **Amoria** : 128, **85**(3, 3a, 3b, 3c)
grossi Iredale 1927, **Volutoconus** : 17, 134, fig. 27c, **89**, 136, 136 fig. 28, **90**(1, 1a)
grossularia Röding 1798, **Drupina** : 92, **61**(5, 5a)
gruneri Reeve 1844, **Vexillum** : 116, **75**(6, 6a)
guntheri Smith 1886, **Paramoria** : 124, **81**(6)
guttata McMichael 1964, **Amoria** : 132, **88**(4)
guttata Röding 1798, **Terebra** : 154, **103**(1, 1a)
Gutturnium Mörch 1852 : 76, 78
Gyrineum Link 1807, : 78

habitat associations
—with alcyonarians : **6, 7**, 42
—with ascidians : 76

—with barnacles : 90
—with bryozoans : **7**
—with fan corals : 60
—with gorgonians : 60, 62
—with hydrozoans : 94
—with mangroves : 83, 84
—with mussels : 88
—with oysters : 88, 92, 96
—with sea anemones : 34, 94
—with sea grasses : 88
—with sea weeds : 98, 118
—with soft corals : 60, 94
—with sponges : **5, 7**, 43, 45, **23, 24**
—with zoanthids : 34
habitat substrates
—coral : 90, 94, 98, 100, 110, 112, 114, 116, 140, 144, 146, 150, 152, 154
—mud : 80, 83, 88, 102, 116, 136
—rocks : 83, 90, 150
—sand : 64, 66, 72, 80, 88, 98, 102, 105, 110, 112, 114, 116, 121, 140, 144, 146, 148, 150, 154, 158
habitat types
—land : 10
—surf : 90, 154, 158
—intertidal : 90, 92, 94, 96, 98, 102, 121, 130, 136, 148
—shallow water : 45, 96, 98, 110, 112, 118, 121, 126, 130, 138, 146, 148, 154
—sublittoral : 90, 92, 94
—continental shelf & slope : 96
—planktonic : 45, 76, 83, 96, 102, 114, 140
—burrowing : 104, 110, 121, 154
—estuarine : 102
Haliotidae : 24
haliotids : 14, 15, 24
Haliotis Linnaeus : 24
halli Cotton 1954 = **Phalium** *pyrum* : 70
hammondae Iredale 1939, **Cypraea** : 56, **35**(9)
hanleyi Angas 1867, **Bedeva** : 88, **59**(12)
hannafordi McCoy 1866, **Pterospira** : 124 fig. 25
hargreavesi Angas 1872, **Volutoconus** : 17, 134, 134 fig. 27a, 136, **90**(5)
Harp shells : 110
Harpa Walch 1771 : 110
harpa Linnaeus 1758, **Harpa** : 110, **72**(3, 3a)
Harpago Mörch 1852, subgen. of **Lambis** : 10, 36
Harpidae : 110
Hastula H. & A. Adams 1853 : 154, 158
hatching of eggs : 83, 96, 105, 121, 140, 148
Haustellum Schumacher 1817 : 83, 84
hectica Linnaeus 1758, **Terebra** : 156, **105**(2)
hedleyi Iredale 1936, **Columbarium** : 96
helenae McMichael 1966 = *mcmichaeli,* subsp. of **Volutoconus** *grossi* : 136
Heliacidae : 34
Heliacus d'Orbigny 1842 : 34
Helmet shells : 66-71
helvola Linnaeus 1758, **Cypraea** : 52, **32**(9)
hepatica Röding 1798, **Septa** : 78, **53**(10, 10a)
herbivorous molluscs : 14, 26, 28, 30, 32, 36
hermaphroditic molluscs : 10, 15
hesitata Iredale 1916, **Cypraea** : 50, **12**(3, 3a, 3b)
Hexaplex Perry 1810 : 82, 84
hilda Iredale 1939, subsp. of **Cypraea** *gracilis* : 56
hippocastanum of Authors *non* Linnaeus, **Thais** : 90
Hipponyx Defrance 1819 : 92, 140
hirundo Linnaeus 1758, **Cypraea** : 58, **38**(4, 4a)
histrio Gmelin 1791, **Cypraea** : 50, **30**(10, 10a)
Homalocantha Mörch 1852 : 84
hoof limpet : 92
howelli Iredale 1931 = **Cypraea** *hesitata* : 50, **30**(3a, 3b)
howensis Iredale 1937 = **Lyria** *deliciosa* : 126
humphreysii Gray 1825, **Cypraea** : 56, **35**(13)
hungerfordi Sowerby 1888, **Cypraea** : 54, **34**(2)
hunteri Iredale 1931, **Cymbiolista** : 122, **80**(5, 5a, 5b)
huttoniae Wright 1879, **Chicoreus** : 86, **58**(10, 10a)

hybrids : 66, 68, 138
hydrozoans : 94
Hypocassis Iredale 1927, subgen. of **Cassis** : 68

Imbricaria Schumacher 1817 : 116
imperata Iredale 1927 = **Phalium** *labiatum* : 70
imperator Woolacott 1956 = **Conus** *parvulus* : 150
imperialis Linnaeus 1758, **Conus** : 16 fig. 8, 142, 144, **97**(7)
imperialis Sowerby 1844, **Epitonium** : 9 fig. 2, 34, **13**(5, 5a)
incarnata Deshayes 1830, **Peristernia** : 100, **65**(10)
incei Philippi 1853, **Polinices** : 64, **45**(4, 4a, 4b)
inermis Angas 1878, **Tudicula** : 112, **73**(6, 6a)
interlirata Reeve 1844, **Cancilla** : 118, **76**(14, 14a)
internal fertilization : 15, 45, 64
intertidal species : 90, 92, 94, 96, 98, 102, 121, 130, 136, 148
invertebrates—as prey : 14
irasans Lamarck 1822, **Oliva** : 106
iredalei Schilder & Schilder 1938, subsp. of **Cypraea** *punctata* : 58
iredalei Abbott 1960, subsp. of **Strombus** *vomer* : 38
irvinae Smith 1909, **Aulicina** : 130, **87**(1)
isabella Linnaeus 1758, **Cypraea** : 50, **30**(7)
ispidula Linnaeus 1758, **Oliva** : 106

jamrachi Gray 1864, **Amoria** : 128, **85**(4, 4a)
jeaniana Cate 1968, form of **Cyraea** *friendii friendii* : 45, **25**(1, 1a)
jourdani Kiener 1839, **Turbo** : 28, **10**(6, 6a)
juveniles : 43, 86, 120, 121, 122, 124, 138, 140, 148 fig. 33

keatsiana Ludbrook 1953, subsp. of **Amoria** *damoni* : 128, **85**(1c, 1d)
kellneri Iredale 1957 = **Aulica** *flavicans* : 132
kenyoniana Brazier 1898, subsp. of **Ericusa** *papillosa* : 122, **80**(2a)
kieneri Tapparone-Canefri 1876, **Cirsotrema** : 34, **13**(8, 8a)
kieneri Hidalgo 1906, **Cypraea** : 58, **38**(5)
kieneri Deshayes 1844, **Thais** : 90, **60**(6, 6a)
kingi Cox 1871 = **Amorena** *sclateri* : 124
klemae Cotton 1953, **Conus** : 148, **99**(3, 3a, 3b)
kreuslerae Angas 1865, **Notovoluta** : 124, **81**(4)

labiatum Perry 1811, **Phalium** : 66, **47**, 70, **49**(5)
labiatus Röding 1798, **Strombus** : 40, **18**(3)
labio Linnaeus 1758, **Monodonta** : 26, **9**(9, 9a)
Labiostrombus Oostingh 1925, subgen. of **Strombus** : 38
labrolineata Gaskoin 1848, **Cypraea** : 52, **32**(6)
Laevicardium Swainson 1840 : 105
Laevistrombus Kira 1955, subgen. of **Strombus** : 40
Lambis Röding 1798 : 10, 36
lambis Linnaeus 1758, **Lambis** : 36, **14**(1), 15
lampas Linnaeus 1758, **Charonia** : 76
Lampusia Schumacher 1817 : 76, 78
lanceata Linnaeus 1767, **Hastula** : 158, **106**(14)
land snails : 10
langfordi Kuroda 1938, **Cypraea** : 54, **34**(10)
larvae : 14, 15, 24, 28, 30, 32, 45, 80, 121
—larval development : 14, 15, 30, 45, 66, 76, 83, 96, 121, 140, 154
—pelagic larvae : 15, 20, 24, 45, 66, 80, 102, 105, 114, 120, 121, 140
—trochophore : 14
—veliger : 14, 15, 15 figs 6 & 7, 66, 76, 105
—larval settlement : 15
laseroni Iredale 1937, **Lyreneta** : 17, 126, **83**(10)
latefasciata Schilder 1930, subsp. of **Cypraea** *asellus* : 56
Latirolagena Harris 1897 : 100

Latirulus Cossmann 1889 : 100
Latirus Montfort 1810 : 98, 100
latiruses : 98, 100
latitudo Garrard 1961, **Tutufa** : 80
lauta Pease 1869, **Hastula** : 158, **106**(7)
lekalekana Ladd 1934, subsp. of **Cypraea** *isabella* : 50
lentiginosus Linnaeus 1758, **Strombus** : 38, **17**(3)
Lentigo Jousseaume 1866, subgen. of **Strombus** : 38
leopardus Röding 1798, **Conus** : 144, **97**(8)
Lepsiella Iredale 1912 : 92
leviathan Schilder & Schilder 1937, **Cypraea** : 50
lignaria Marrat 1868, **Oliva** : 106, **69**(2, 2a, 2b)
lima Deshayes 1859, **Terebra** : 156, **105**(4)
limacina Lamarck 1810, **Cypraea** : 58, **38**(1)
limpet : 14, 15
lineata Gmelin 1791, **Nerita** : 30, **11**(7, 7a)
lineata Leach 1814 = **Zebramoria** *lineatiana* : 132
lineatiana Weaver & Du Pont 1967, **Zebramoria** : 132, **88**(5, 5a, 5b, 5c)
lineatus Lamarck 1822, **Trochus** : 26, **9**(13, 13a)
lineolata Lamarck 1809, **Cominella** : 96, **63**(8, 8a, 8b, 8c)
litterata Lamarck 1811, **Strigatella** : 118, **76**(9)
litteratus Linnaeus 1758, **Conus** : 142, 144, **97**(9) ·
littorea Linnaeus 1758, **Littorina** : 15 fig. 6
Littorina Férussac 1821 : 15 fig. 6
Littorinidae : 30
lividus Hwass 1792, **Conus** : 142, 150, **100**(12)
Livonia Gray 1855 : 122
lobata Blainville 1832, **Drupina** : 92, **61**(4)
Lobophytum Marenzeller 1886 : 6
locomotion—in sand : 36, 105, 114, 121
—swimming : 105
longior Iredale 1935, subsp. of **Cypraea** *caurica* : 52
lotorium Linnaeus 1758, **Cymatium** : 76, **52**(4)
lugubris Swainson 1822, **Mitra** : 116, **75**(2)
luhuanus Linnaeus 1758, **Strombus** : 38, **17**(5, 5a)
lutea Gmelin 1791, **Cypraea** : 56, **35**(1, 1a)
lyncichroa Melvill 1888, subsp. of **Cypraea** *tigris* : 48
lynx Linnaeus 1758, **Cypraea** : 50, **30**(8, 8a)
Lyreneta Iredale 1937 : 17, 126
Lyria Gray 1847 : 126

macandrewi Sowerby 1887, **Amoria** : 121, 128, **84**, **85**(6, 6a)
macgillivrayi Dohrn 1862, **Murex** : 84, **57**(5)
macula Angas 1867, subsp. of **Cypraea** *gracilis* : 56, **35**(12, 12a)
maculata Swainson 1822, **Amoria** : 130, **86**, **87**(7, 7a, 7b)
maculata Linnaeus 1758, **Terebra** : 154, **103**(6, 6a)
maculatus Linnaeus 1758, **Trochus** : 26, **9**(11, 11a)
maculosa Gmelin 1791, **Colubraria** : 54(9, 9a), 82
madreporarum Sowerby 1832, **Quoyula** : 94, **62**(9)
Magilidae : 94
magnifica Gebauer 1802, **Cymbiolena** : 17, 122, **79**, **80**(1)
magus Linnaeus 1758, **Conus** : 152, **101**(14, 14a)
major Röding 1798, **Harpa** : 110, **72**(2)
Malea Valenciennes 1833 : 74
Mamillaria Swainson 1840 : 64
Mamillinae, subfam. of Naticidae : 64
mammilla Sowerby 1844, **Livonia** : 122, **80**(3), 138
Mammilla Schumacher 1817 : 64
Mancinella Link 1807 : 90
mancinella Linnaeus 1758, **Mancinella** : 90, **61**(2, 2a)
mangrove dwellers : 83, 84

mappa Linnaeus 1758, **Cypraea** : 48, **27**, **28**(6)
mara Iredale 1931 = **Cypraea** *eburnea* = **C.** *miliaris* : 52
marcia Iredale 1939, subsp. of **Cypraea** *ursellus* : 58
margariticola Broderip 1832, **Morula** : 92, **61**(9)
marginalba Blainville 1832, **Morula** : 92, **61**(8)
marginata Lamarck 1811, **Alocospira** : 108, **71**(5)
marginata Gaskoin 1848, **Cypraea** : 42, 48, **28**(2, 2a)
Marmarostoma Swainson 1829, subgen. of **Turbo** : 28
marmoreus Linnaeus 1758, **Conus** : 144, **97**(6)
Maugean province : 18
mauritiana Linnaeus 1758, **Cypraea** : 48, **28**(9)
mawsoni Cotton 1948 = **Phalium** *pyrum* : 70
maxima Philippi 1848, **Architectonica** : 34, **13**(12, 12a)
maximus Tryon 1881, **Penion** : 96, **63**(1, 1a)
Mayena Iredale 1917 : 76
mcmichaeli Habe & Kosuge 1966, subsp. of **Volutoconus** *grossi* : 134, 136, **90**(2)
Megalatractus Fischer 1884 : 98
melaniana Lamarck 1811 = **Eumitra** *nigra* : 118
melanotragus Smith 1884 = **Nerita** *atramentosa* : 30
Melaraphe Menke 1828 : 17, 30
melosus Hedley 1924, **Polinices** : 64, **45**(2, 2a)
Melo Sowerby 1826 : 120, 121, 138
Melon shells : 138
Melongenidae : 96
melvilli Hidalgo 1906, subsp. of **Cypraea** *felina* : 56
melwardi Iredale 1930 = **Cypraea** *cribraria* : 52, **32**(10a)
merces Iredale 1924, **Ellatrivia** : 62, **44**(12)
Mesericusa Iredale 1929 : 122, 138,
Mesogastropoda : 10, 12 fig. 4, 14, 15
metavona Iredale 1935, subsp. of **Cypraea** *miliaris* : 52
microdon Gray 1828, **Cypraea** : 56, **35**(11)
miles Linnaeus 1758, **Conus** : 146, **98**(9, 9a)
miliaris Gmelin 1791, **Cypraea** : 52, **32**(8, 8a)
miliaris Gmelin 1791, **Vitularia** : 88, **59**(11)
Millipes Mörch 1853, subgen. of **Lambis** : 36
miltonis Gray 1834, **Melo** : 17, 120, 138, **92**(3)
miniacea Röding 1798, **Oliva** : 67, 106, **69**(8, 8a)
minor Lamarck 1816 = **Harpa** *amouretta* : 110
Mitra Röding 1798 : 114
mitra Linnaeus 1758, **Mitra** : 114, **74**(1, 1a)
mitraeformis Lamarck 1811, **Lyria** : 126, **83**(7, 7a, 7b)
mitratus Hwass 1792, **Conus** : 152, **101**(7, 7a)
Mitres : 114-117, 140
Mitridae : 114-117
molleri Iredale 1931, **Cypraea** : 43, 58, **38**(8)
molleri Iredale 1936, **Relegamoria** : 132, **88**(9, 9a)
molluscs as food : 14, 76, 83, 90, 96, 98, 140, 142
monachus Linnaeus 1758, **Conus** : 152, **101**(9, 9a, 9b)
moneta Linnaeus 1758, **Cypraea** : 56, **35**(2)
monile Reeve 1863, **Calliostoma** : 26, **9**(6, 6a)
monilifera Reeve 1864, **Alocospira** : 108, **71**(6, 6a)
monodon Sowerby 1825 = **Murex** *cornucervi* : 86
Monodonta Lamarck 1799 : 26
Monoplacophora : 9
Monoplex Perry 1811 : 76
Moon shells : 64
Mopsella Gray 1858 : 37
moretensenae Iredale 1957, **Grandeliacus** : 34, **13**(14, 14a)
moretonensis Schilder 1965, subsp. of **Cypraea** *langfordi* : 54, **34**(10)
Morula Schumacher 1817 : 92
morum Röding 1798, **Drupa** : 92, **61**(2)
Mudwhelks : 32
Mulberry whelk : 92

multiplicatum Sowerby 1895, **Haustellum** : 84, **57**(7, 7a)
Murex Linnaeus 1758 : 82, 84
Murex shells : 14, 15, 82-89, 94, 96
Murexsul Iredale 1915 : 88
Muricidae : 82-90, 94, 96
Muricinae : 82, 83
muricinum Röding 1798, **Gutturnium** : 76, 78, **53**(4, 4a)
muscaria Lamarck 1822 = **Terebra** *areolata* : 154
mussel beds : 88
mustellina Lamarck 1811, **Oliva** : 108, **71**(2, 2a, 2b)
mutabilis Swainson 1821, **Strombus** : 36, 40, **18**(4)

nana Tenison Woods 1879, **Cassis** : 68, **48**(2, 2a)
Nannamoria Iredale 1929 : 132
Naquetia Jousseaume 1808, subgen. of **Pterynotus** : 83, 84
nashi Iredale 1931, subsp. of **Cypraea** *labrolineata* : 52
Nassa Röding 1798 : 94
Nassariidae : 102
Nassarius Duméril 1806 : 102
nassatula Lamarck 1822, **Peristernia** : 100, **65**(9)
Natica Scopoli 1777 : 64
Naticarius Duméril 1806 : 64
Naticas : 64
Naticidae : 64
Naticinae, subfam. of Naticidae : 64
nautilus : 9
nebulosa Sowerby 1825, **Terebra** : 156, **105**(7)
nectarea Iredale 1930, **Volva** : **41**
neglecta Adams & Reeve 1850, **Epitonium** : 34
Negyrina Iredale 1929 : 78
Neocancilla Cernohorsky 1966 : 114
Neogastropoda : 10, 14, 15
Neosimnia Fischer 1884 : 60, 62
Nerita Linnaeus 1758 : 30
Nerites : 30
Neritidae : 30
neurotoxin : 142
Nevia Jousseaume 1887 : 108
New Zealand fauna : 20, 90, 96, 154
nicobaricum Röding 1798, **Lampusia** : 76, 78, **53**(3)
nicobaricus Lamarck 1822, **Fusinus** : 98, **64**(8, 8a)
nielsenae Marsh 1962, **Conus** : 152, **101**(5)
nielseni McMichael 1959, subsp. of **Cymbiolacca** *complexa* : 126, **83**(5a)
nigra Gmelin 1791, **Eumitra** : 118, **76**(4)
nigrospinosus Reeve 1845, **Murex** : 84, **57**(1)
nigrosulcata Reeve 1855, **Patelloida** : 16 fig. 8
Ninella Gray 1850 : 28
Niotha H. & A. Adams 1853 : 20, 102
nitida Hinds 1844, **Hastula** : 158, **106**(13)
niveum Brazier 1872, form of **Phalium** *pyrum* : 70, **49**(1)
nivosa Lamarck 1804, **Aulicina** : 17, 130, **87**(2, 2a)
nobilis Röding 1798 = **Harpa** *harpa* : 110
Nodilittorina von Martens 1897 : 30
nodiplicata Cox 1910, **Cottonia** : 124, 124 fig. 24, **81**(1)
nodostaminea Hedley 1912, **Cancilla** : 118, **76**(12)
nodulosum Bruguière 1792, **Cerithium** : 32, **12**(5, 5a)
nodulosus Sowerby 1864, **Conus** : 17, 146, **98**(6)
norrissii Gray 1838, **Voluta** : 130
Northern Australian Region : 17, 18 fig. 10
Notocochlis Powell 1933 : 64
Notocypraea Schilder 1927, subgen. of **Cypraea** : 43, 58
Notopeplum Finlay 1927 : 122 fig. 23
Nototerebra Cotton 1947 : 158
Notovoluta Cotton 1946 : 124

noumeensis Marie 1869, subsp. of **Cypraea** *annulus* : 56
novaehollandiae A. Adams 1854, ? subsp. of **Conus** *anemone* : 148, **99**(8, 8a, 8b)
novaehollandiae Reeve 1848, **Fusinus** : 6 fig. 1, 98, **64**(5)
nucea Gmelin 1791, **Pterygia** : 114, **74**(7, 7a)
nucleus Linnaeus 1758, **Cypraea** : 58, **36**, **38**(3, 3a)
nucleus Lamarck 1811, **Lyria** : 126, **83**(9)
nudibranchs : 10, 14
nugata Iredale 1935, subsp. of **Cypraea** *saulae* : 56
nurse eggs : 15, 66, 96
nussatella Linnaeus 1758, **Conus** : 150, **100**(8)

obeliscus Gmelin 1791 = **Trochus** *pyramis* : 26
oblita Smith 1909, subsp. of **Aulicina** *nivosa* : 130, **87**(3, 3a)
obscurus Sowerby 1833, **Conus** : 142, 150, **100**(1)
occidentalis Iredale 1935, ? subsp. of **Cypraea** *declivis* : 58
occidua Cotton 1946, **Notovoluta** : 124
Octopus : 9
offlexa Iredale 1931 = **Architectonica** *reevei* : 34
oligostira Tate 1891, **Penion** : 96, **63**(3)
Oliva sp. : 106, **69**(1, 1a)
oliva Linnaeus 1758, **Oliva** : 106, **69**(10, 10a, 10b)
Oliva Bruguière 1789 : 104-109
Olivellinae : 104
Olives : 104-109
Olividae : 104-109
Olivinae : 104-109
omaria Hwass 1792, **Conus** : 142, 146, 146 fig. 32, **98**(2, 2a)
oncus Röding 1798, **Naticarius** : 64, **45**(11, 11a)
opaca Récluz 1851, **Mammilla** : 64, **45**(6, 6a)
Opalia H. & A. Adams 1853 : 34
Opisthobranchia : 10, 14
opposita Iredale 1937 = **Lyria** *deliciosa* : 126
orbita Gmelin 1791, **Dicathais** : 15 fig. 6, 20, 90, **60**(10)
orca Cotton 1951 = **Ericusa** *fulgetrum* : 124
orcina Iredale 1931, subsp. of **Cypraea** *vitellus* : 48
ornata Marrat 1867 = **Oliva** *lignaria* : 106
orrae Abbott 1960, subsp. of **Strombus** *urceus* : 40
oryza Lamarck 1810, **Trivirostra** : 62, **43**, **44**(11)
ovina Gmelin 1791, **Haliotis** : 24, **8**(10, 10a)
Ovula Bruguière 1789 : 60, 62
Ovulidae : **1**, 15 fig. 6, 60-63
ovulids : **1**, 15 fig. 6, 60-63
ovum Gmelin 1791, **Cypraea** : 52, **32**(4, 4a)
ovum Linnaeus 1758, **Ovula** 60, 42, 62, **44**(6)
oysters : 88, 92, 96

paetelianus Kobelt 1876, **Latirus** : 100, **65**(6)
Pagoda shells : **62**(10, 10a), 96
pagodus Linnaeus 1758, **Tectarius** : 30, **11**(9)
pale shelled forms : 46, 58
palinodium Iredale 1931 = **Phalium** *thomsoni* : 70
pallasi Kiener 1838-39, **Epitonium** : 34, **13**(4, 4a, 4b)
pallidula Gaskoin 1849, **Cypraea** : 56, **35**(4)
papalis Linnaeus 1758, **Mitra** : 114, **74**(2, 2a)
papilio Link 1807, **Neocancilla** : 114, **74**(9, 9a)
papilliferus Sowerby 1834, **Conus** : 10, 140, **95**, 148, 148 fig. 33, **99**(10)
papillosa Swainson 1822, **Ericusa** : 122, **80**(2, 2a)
papillosus Linnaeus 1758, **Alectrion** : 102, **66**(2, 2a)
parabola Garrard 1960, **Nannamoria** : 132, **88**(7, 7a)
Paramoria McMichael 1960 : 124
parasites—flat worm : 32
—molluscs : 14
Parcanassa Iredale 1936 : 102

pardalis Shaw 1795, subsp. of **Cypraea** *tigris* : 48

parkinsoniana Perry 1811, **Austrosassia** : 78, **53**(9, 9a)

parthenopeum von Salis 1793, **Cymatium** : 76

particeps Hedley 1915, **Alectrion** : 11 fig. 3, 102

parvulus Link 1807, **Conus** : 150, **100**(14)

patagiatus Hedley 1912, **Pterynotus** : 88, **59**(6)

Patelloida Quoy & Gaimard 1834 : 16 fig. 8

patriarchalis Gmelin 1791, **Pusia** : 118, **76**(8)

pattersoniana Perry 1811 = **Lyria** *nucleus* : 126

patula Pennant 1777, **Simnia** : 15 fig. 6

pauciruge Menke 1843, **Phalium** : 70, **49**(2, 2a)

pauperata Lamarck 1822, **Parcanassa** : 102, **66**(5, 5a)

pecten Solander, 1786, **Murex** : 83, 84, **57**(2)

pelagic larvae : 15, 20, 24, 45, 80

pele Pilsbry 1921, **Homalocantha** : 84

pelecypods : 9

Pencil shells : 154-159

Penion Fischer 1884 : 96

pentella Iredale 1939, subsp. of **Cypraea** *teres* : 52

percomis Iredale 1931 = **Cypraea** *cernica tomlini* : 52

perconfusa Iredale 1935, subsp. of **Cypraea** *eglantina* : 50

percum Perry 1811, **Gyrineum** : 78

perdix Hinds 1844, **Architectonica** : 34, **13**(9, 9a)

perdix Wood 1828 = **Phasianella** *ventricosa* : 28

perdix Linnaeus 1758, **Tonna** : **50**, 74, 74 fig. 16, **51**(4, 4a)

Peristernia Mörch 1852 : 98, 100

peristernias : 98, 100

peristicta McMichael 1963, **Cymbiolacca** : 126, **83**(3, 3a)

Periwinkles : 30

permaestus Hedley 1915, **Pterynotus** : 83, 84, **57**(8)

Peronian province : 18

peroniana Iredale 1940 = **Lyria** *nucleus* : 126

peronianus Iredale 1931 = **Conus** *anemone* : 148

perplexa Pease 1860, **Epitonium** : 34, **13**(3, 3a)

perplicata Hedley 1902, **Notovoluta** : 124, 132, **88**(3)

perryi Iredale 1912, subsp. of **Casmaria** *ponderosa* : 70

perryi Ostergaard & Summers 1957 = **Cymbiolacca** *pulchra* : 126, **83**(4b)

persica Linnaeus 1758, **Purpura** : 94, **62**(1)

perspectiva Linnaeus 1758, **Architectonica** : 34, **13**(11, 11a, 11b)

pertusa Born 1778, **Terebra** : 158, **106**(2)

petholatus Linnaeus 1758, **Turbo** : 28, **10**(7, 7a)

Phalium Link 1807 : 7 fig. 2, 66, 68, 70

Phasianella Lamarck 1804 : 28

Phasianotrochus Fischer 1885 : 26

philippensis Watson 1886, **Typhis** : 88, **59**(10, 10a)

Philippia Gray 1847 : 34

philippinarum Sowerby 1848, **Volva** : 16 fig. 8, 62, **44**(8)

Phos Montfort 1810 : 96

phytoplankton—as food : 15

pica Blainville 1832 = **Mancinella** *tuberosa* : 90

pictus Reeve 1847, **Latirus** : 100, **65**(4)

pileare Linnaeus 1758, **Lampusia** : 76, 78, **53**(12, 12a)

pileola Reeve 1842, **Astraea** : 28, **10**(8, 8a)

piperita Gray 1825, **Cypraea** : 43, 58, **37, 38**(13, 13a)

pipus Röding 1798, **Strombus** : 38, **17**(4)

plankton—as food : 14, 94
—larval stage : 45, 76, 83, 96, 102, 114, 140

planktonic larvae : 102, 114, 140

planorbis Born 1778, **Conus** : 148, **99**(14)

Pleuroploca Fischer 1884 : 98

Plicarcularia Thiele 1929 : 102

plicarium Linnaeus 1758, **Vexillum** : 116, **75**(7, 7a)

plicata Linnaeus 1758, **Nerita** : 30, **11**(5, 5a)

plicatus Röding 1798, **Strombus** : 40, **18**(6, 6a)

Pocillopora Lamarck 1816 : 94

Polinices Montfort 1810 : 64

polita Linnaeus 1758, **Nerita** : 30, **11**(1, 1a)

polygonus Gmelin 1791, **Latirus** : 100, **65**(2)

pomum Linnaeus 1758, **Malea** : 74, **51**(7)

ponderosa Gmelin 1791, **Casmaria** : 70, **49**(6)

pontificalis Lamarck 1810 = **Conus** *dorreensis* : 148

poraria Linnaeus 1758, **Cypraea** : 54, **34**(7, 7a)

Potamididae : 32

powelli Cotton 1956, **Charonia** : 76, **52**(3)

powisiana Récluz 1844, **Mamillaria** : 64, **45**(5, 5a)

praetexta Reeve 1849, **Amoria** : 17, 128, **85**(2, 2a)

predation : 14, 20, 43, 64, 66, 72, 76, 90, 96, 105, 114, 120, 121, 140, 142, 154
—by cassids : 66
—by volutes : 120
—on Crown of Thorns starfish : 20, 76
—on echinoids : 66

pricei Smith 1887, **Saginafusus** : 98, **64**(1)

Primovula Thiele 1925 : **40**

prodiga Iredale 1939 = **Cypraea** *cernica tomlini* : 52

Propefusus Iredale 1924 : 98

propinqua Tenison Woods 1876, subsp. of **Lepsiella** *vinosa* : 92

propodium of ancillid : 104

Prosobranchia : 10, 12, 14

protection : 83, 121

provocationis McMichael 1961, **Pseudocymbiola** : 126, **83**(6)

pseudamygdala Hedley 1903, **Cronia** : **61**(13), 94

Pseudocymbiola McMichael 1961 : 126

Pterospira Harris 1897 : 17, 124 fig. 25

Pterygia Röding 1798 : 114

Pterynotus Swainson 1833 : 88

pulchellum Forbes 1852, **Gyrineum** : 78, **53**(15, 15a)

pulchellus Reeve 1851, subsp. of **Strombus** *plicatus* : 40, **18**(6, 6a)

pulcher Reeve 1842, **Turbo** : 16 fig. 8, 27, **10**(3)

pulchra Sowerby 1825, **Cymbiolacca** : 17, 121, **82, 83**(4, 4a, 4b)

pulicaria Reeve 1846, **Cypraea** : 43, 58, **38**(11, 11a)

pulicarius Hwass 1792, **Conus** : 142, 150, **100**(7)

Pulmonata (pulmonates) : 11, 12, 14

punctata Verco 1906, **Austroharpa** : 110, **72**(5, 5a)

punctata Linnaeus 1771, **Cypraea** : **22**, 58, **38**(7, 7a)

punctata Duclos 1831, **Primovula** : **40**

purissima Vredenburg 1919, subsp. of **Cypraea** *erosa* : 52

Purples : 90-95

Purpura Bruguière 1792 : 94

Pusia Swainson 1840 : 118

Pustularia Swainson 1840, subgen. of **Cypraea** : 62

pyramidalis Quoy & Gaimard 1833, **Nodilittorina** : 30, **11**(11, 11a)

pyramis Born 1778, **Trochus** : 16 fig. 8, 26, **9**(12, 12a)

Pyrazus Montfort 1810 : 32

pyriformis Gray 1824, **Cypraea** : 54, **34**(11, 11a)

pyrum Linnaeus 1758, **Ranularia** : 76, **52**(5)

pyriformis Récluz 1844, **Polinices** : 64, **45**(3, 3a)

pyrrhus Menke 1843, **Niotha** : 20, 102, **66**(4, 4a)

pyrulatus Reeve 1847, **Propefusus** : 98, **64**(4, 4a)

pyrum Lamarck 1822, **Phalium** : 66, 68, 70, **49**(1, 1a, 1b, 1c)

quadrimaculata Gray 1924, **Cypraea** : 56, **35**(7)

quaesita Iredale 1956 = **Aulica** *flavicans* : 132

queenslandica Schilder 1966, **Cypraea** : 54, **34**(3)

quercinus Solander 1786, **Conus** : 142, 144, **97**(11)

Quoyula Iredale 1912 : 94

radiata Röding 1798, **Philippia** : 34, **13**(13, 13a)

radula, & radular teeth : 14, 14 fig. 5, 45, 136 fig. 28, 141 fig. 30b

ramosus Linnaeus 1758, **Chicoreus** : 86, **58**(1)

rana Linnaeus 1758, **Bursa** : 80, **54**(6, 6a)

randalli Stokes 1961, form of **Cymbiolacca** *wisemani* : 126, **83**(1a)

Ranella Lamarck 1816 : 80

Ranularia Schumacher 1817 : 76

Rapa Bruguière 1792 : 94

rapa Linnaeus 1758, **Rapa** : 94, **62**(5, 5a)

rasilistoma Abbott 1959, **Tudicula** : 112, **73**(7, 7a)

rattus Hwass 1792, **Conus** : 152, **101**(13)

razor clam : 120

recticornis von Martens 1880, **Chicoreus** : 86, **58**(8)

recurvirostris Schubert & Wagner 1827, **Latirus** : 100, **65**(3)

reevei Sowerby 1864, subsp. of **Amoria** *damoni* : 128, **85**(1)

reevei Hanley 1862, **Architectonica** : 34, **13**(10, 10a)

reevei Sowerby 1832, **Cypraea** : 43, 50, **30**(4, 4a)

regina Gmelin 1791, subsp. of **Cypraea** *mauritiana* : 48

Relegamoria Iredale 1936 : 132

reproduction—maturity : 17, 43
—reproductive system : 10, 14, 15, 154
—breeding habits : 66, 72, 105, 114, 140, 154
—copulation : 15, 45
—spawning : 66, 83, 140
—sperm : 14, 15, 24
—fertilization—external : 14, 24, 26
—internal : 154
—eggs & egg masses : 14, 15, **6**, 24, 26, 28, 30, 32, 36, 45, **26**, 64, 66, 96, 120, 121, 122, 140, 154
—egg capsules : **4,** 45, 66, **47,** 76, 80, 82 fig. 17, 83, 96, 102, 105, 114, 120, 121, 140, **95,** 144 fig. 31, 148, 154
—nurse eggs : 15, 66, 96
—embryos : 15 fig. 7, 83, 105, 120, 121, 140
—hatching : 83, 96, 105, 121, 148

reticulata Reeve 1844 = **Amoria** *damoni* : 128

reticulata Röding 1798, **Distorsio** : 78, **53**(8, 8a)

reticulata Blainville 1832, **Lepsiella** : 92

reticulata Röding 1798, **Oliva** : 106, **69**(7, 7a)

reticulifera Schilder 1924 = **Cypraea** *piperita* : 58

retiolus Hedley 1914, **Fusitriton** : 78, **53**(2)

rhinoceros Sowerby 1865, subsp. of **Cypraea** *pallidula* : 56

Rhinoclavis Swainson 1840 : 32

rhodia Reeve 1845 = **Eumitra** *badia* : 118

rhodostoma Sowerby 1835, **Bursa** : 80

rhomboides Schilder & Schilder 1933, subsp. of **Cypraea** *moneta* : 56

ricinus Linnaeus 1758, **Drupa** : 92, **61**(3, 3a)

Rimella Agassiz 1840 : 36, 40

roadnightae McCoy 1881, **Livonia** : 122, **80**(4)

roei Gray 1826, **Haliotis** : 24, **8**(12)

rosa Perry 1811, **Bursa** : 80, **54**(4, 4a)

rosselli Cotton 1948, **Cypraea** : 17, 42, 48, **28**(3)

rota Mawe 1823, **Murex** : 84

rotularia Lamarck 1822, **Astraea** : 28, **10**(9, 9a)

rubecula Linnaeus 1758, **Septa** : 78, **53**(11, 11a)

ruber Leach 1814, **Haliotis** : 24, **8**(1)

rubeta Röding 1798, **Tutufa** : 80

rubicunda Perry 1811, **Charonia** : 20, 76, **52**(2)

rubiginosus Reeve 1845, **Chicoreus** : 86, **58**(4, 4a, 4b)

rubusidaeus Röding 1798, **Drupa** : 92, **61**(1, 1a)

rudis Gray 1826, **Austrocochlea** : 26, **9**(7, 7a)

rufa Linnaeus 1758, **Cypraecassis** : 66, **46**(2, 2a)

trigonus Reeve 1848, **Conus** : 146, **98**(11, 11a)
tripterus Born 1778, **Pterynotus** : 88, **59**(7)
triqueter Born 1778, **Pterynotus** : 88, **59**(5, 5a)
triremis Perry 1811 = **Murex** *pecten* : 83, 84
triseriata Gray 1834, **Terebra** : 156, **104, 105**(8)
Tritonaliinae : 83
tritoniformis Blainville 1832, **Agnewia** : 20
tritonis Linnaeus 1758, **Charonia** : 20, 21, 76, **52**(1)
Tritons : 76-79
Trivias : 62
Triviidae : 62
Trivirostra Jousseaume 1884 : 62
Trochidae : 14, 26
trochophore : 14
Trochus Linnaeus 1758 : 26
Trumpet shells : 20, 76-79
truncata Humphrey 1786, **Lambis** : 36, **14**(2)
tuberosa Röding 1798, **Mancinella** : 90, **60**(3)
Tudicula H. & A. Adams 1863 : 112
tulipa Linnaeus 1758, **Conus** : 142, 144, **97**(2)
Tun shells : 72-75
Turban shells : 14, 28
turbinellum Linnaeus 1758, **Vasum** : 112, **73**(2, 2a)
Turbinidae : 14, 28
Turbo Linnaeus 1758 : 28
turneri Gray 1834, **Amoria** : 128, 130, **87**(5)
Turritellidae : 154
Turritriton Dall 1904 : 78
turritus Gmelin 1791, **Latirulus** : 100, **65**(5)
Tutufa Jousseaume 1881 : 80
tweedianum Macpherson 1962, **Haustellum** : **55**, 84, **57**(6, 6a)
Typhinae : 88
Typhis Montfort 1810 : 88
tyria Reeve 1843, **Angaria** : 26, **9**(1, 1a)

Umbilia Jousseaume 1884, subgen. of **Cypraea** : 43, 50
umbilicata Quoy & Gaimard 1833, **Sigaretotrema** : 64, **45**(13, 13a)
umbilicatus Sowerby 1826, **Melo** : 138, **92**(1, 1a)
umbilicatus Tenison Woods 1875, **Murexsul** : 88, **28**(9, 9a)
undata Linnaeus 1758, **Nerita** : 30, **11**(4, 4a, 4b)
undatus Lamarck 1816, **Clanculus** : 26, **9**(3, 3a)
undosus Linnaeus 1758, **Cantharus** : 96, **63**(6, 6a)
undulata Lamarck 1804, **Amorena** : 124, **81**(8, 8a), 128
undulata Solander 1786, **Subninella** : 28, **10**(4, 4a)
undulata Gray 1834, **Terebra** : 158, **106**(12)
unifasciata Gray 1826, **Melaraphe** : 17, 30, **11**(12, 12a)
urceus Linnaeus 1758, **Strombus** : 40, **18**(2, 2a, 2b)
Urchins : 66, 72
ursellus Gmelin 1791, **Cypraea** : 58, **38**(6)
uva Röding 1798, **Morula** : 92, **61**(7, 7a)

vanelli Linnaeus 1758, subsp. of **Cypraea** *lynx* : 50
varia Linnaeus 1758, **Haliotis** : 24, **8**(11, 11a)

variabilis Swainson 1820, **Strombus** : 40, 40 fig. 11, **18**(8, 8a)
variabilis Reeve 1844, **Mitra** : 116, **75**(4, 4a)
variations in shell form :
—clines : 45, 46, 88
—geographic variants : 45
—hybrids : 66, 68, 138
—colour forms : 46, 58, 104, 122, 124, 128, 144, 146, 148, 150
—dwarf forms : 43, 45, 46, 98, 118, 121, 122, 124, 128, 136, 138, 146
—aberrant forms : 45, 46
varicosa Sowerby 1841, **Homolocantha** : 84
varicosa Lamarck 1822, **Cirsotrema** : 34, **13**(7, 7a)
variegata Röding 1798 = **Oliva** *reticulata* : 106
variegata Lamarck 1822, **Tonna** : 74, **51**(1, 1a)
variegata Gmelin 1791, **Torinia** : 34, **13**(15, 15a, 15b)
Vase shells : 112
Vasidae : 112
Vasum Röding 1798 : 112
vaticina Iredale 1931, subsp. of **Cypraea** *subviridis* : 54
Velacumantus Iredale 1936 : 32
velesiana Iredale 1936, **Ancillista** : 108, **71**(4)
Veliger : 14, 15, 15 fig. 6 & 7, 66, 76, 105
venomous cone shells : 140, 141, 142, 144, 146, 148, 150, 152, 154
—venom : 140, 142, 143, 144
—venom apparatus : 140, 141 fig. 30a, 142
—treatment : 143
ventricosa Swainson 1822, **Phasianella** : 28, **10**(10)
venusta Sowerby 1847, **Cypraea** : 42, 46, **23, 24, 25**(3, 3a, 3b, 3c)
vercoi Schilder 1930, form of **Cypraea** *friendii friendii* : 45, **25** fig. 1c
verconis Cotton & Godfrey 1932 = **Cypraea** *angustata* : 58
verconis Tate 1892, **Notovoluta** : 124, **81**(3)
verrucosus Linnaeus 1758, **Calpurnus** : **6**, 62, **44**(5, 5a)
vertagus Linnaeus 1758, **Rhinoclavis** : 32, **12**(4, 4a, 4b)
Vexilla Swainson 1884 : 94
vexillum Sowerby 1835, **Argobuccinum** : 78, **53**(1)
vexillum Gmelin 1791, **Conus** : 144, **97**(10)
vexillum Gmelin 1791, **Vexilla** : 94, **62**(2)
Vexillum Röding 1798 : 116
victoriae Reeve 1843, **Conus** : 17, 146, **98**(5, 5a, 5b, 5c, 5d)
vidua Röding 1798, **Oliva** : 106
violacea Kiener 1836, **Coralliophila** : 94, **62**(8)
vinosa Lamarck 1822, **Lepsiella** : 92, **61**(10, 10a)
viridicolor Cate 1962, subsp. of **Cypraea** *cernica* : 52, **32**(12)
virgo Linnaeus 1758, **Conus** : 144, **97**(3)
vitellus Linnaeus 1758, **Cypraea** : 43, fig. 12, 48, **28**(8)
vitiliginea Menke 1843, **Ethminolia** : 16 fig. 8
vittata Deshayes 1831, subsp. of **Cypraea** *ziczac* : 56
vittatus Linnaeus 1758, **Strombus** : 38, **17**(8), 40
Vitularia Swainson 1840 : 88
vitulinus Hwass 1792, **Conus** : 152, **101**(6)

vixlirata Cotton 1943 = **Haliotis** *conicopora* : 24
Volegalea Iredale 1938 : 98
Volutes : 14, 120-139
Volutidae : 120-139
Volutoconus Crosse 1871 : 17, 134-137
Volva Röding 1798 : 16 fig. 8, 60, 62
volva Gmelin 1791, ? = **Amoria** *maculata* : 130
volva Linnaeus 1758, **Volva** : **39**, 62, **44**(4, 4a)
vomer Röding 1798, **Strombus** : 38, **17**(7, 7a)
vulpecula Linnaeus 1758, **Vexillum** : 116, **75**(11, 11a, 11b)

waitei Hedley 1903, **Penion** : 96, **63**(2)
walkeri Sowerby 1832, **Cypraea** : 54, **34**(6, 6a)
walkeri Melvill 1895, **Latirus** : 100, **65**(8)
wardiana Iredale 1938, **Volegalea** : **63**(9), 98
waterhousei Adams & Angas 1864, **Cabestana** : 78, **53**(13, 13a)
Wentletraps : 34
weaveri McMichael 1961, **Paramoria** : 124
Western Overlap Zone : 17, 18, 18 fig. 10
westralis Iredale 1935, subsp. of **Cypraea** *histrio* : 50
whelks : 90
—busycon whelks : 96
—dog whelks : 102
whitleyi Iredale 1939, subsp. of **Cypraea** *clandestina* : 56
whitleyi Iredale 1949 = **Ninella** *torquata* : 28
whitworthi Cate 1964, subsp. of **Cypraea** *chinensis* : 52
whitworthi Abbott 1968, **Phalium** : 70, **49**(4, 4a)
wilkinsi Griffiths 1959, **Cypraea** : 43
wilsoni Abbott 1967, **Strombus** : 40
wisemani Brazier 1870, **Cymbiolacca** : **3**, 126, **83**(1, 1a)
woolacottae McMichael 1958 = **Cymbiolacca** *pulchra* : 126, **83**(4a)
worms—as food : 14, 80, 140, 154

Xancidae : 112
xanthodon Sowerby 1832, **Cypraea** : 17, 54, **34**(9, 9a)
Xenogalea Iredale 1927 = **Phalium** : 70
Xenophalium Iredale 1927, subgen. of **Phalium** : 66

yatesi Crosse 1865, **Typhis** : 88

zebra Leach 1814, **Zebramoria** : **78**, 128, 132, **88**(6)
Zebramoria Iredale 1929 : 121, 132
Zemira H. & A. Adams 1858 : 108
Zeuxis H. & A. Adams 1853 : 102
ziczac Linnaeus 1758, **Cypraea** : 56, **35**(14)
Zoanthids : 34
Zoila Jousseaume 1884, subgen. of **Cypraea** : 42, 43, 45-48
zonale Quoy & Gaimard 1833, **Ectosinum** : 64, **45**(14, 14a)